ELECTRICITY

TASK CARD SERIES

Conceived and
written by
Ron Marson

Illustrated by
Peg Marson

TOPS LEARNING SYSTEMS

10970 S. Mulino Rd.
Canby OR 97013

Oh, those pesky COPYRIGHT RESTRICTIONS *!*

Dear Educator,

TOPS is a nonprofit organization dedicated to educational ideals, not our bottom line. We have invested much time, energy, money, and love to bring you this excellent teaching resource.

And we have carefully designed this book to run on simple materials you already have or can easily purchase. If you consider the depth and quality of this curriculum amortized over years of teaching, it is dirt cheap, orders of magnitude less than prepackaged kits and textbooks.

Please honor our copyright restrictions. We are a very small company, and book sales are our life blood. When you buy this book and use it for your own teaching, you sustain our publishing effort. If you give or "loan" this book or copies of our lessons to other teachers, with no compensation to TOPS, you squeeze us financially, and may drive us out of business. Our well-being rests in your hands.

What if you are excited about the terrific ideas in this book, and want to share them with your colleagues? What if the teacher down the hall, or your homeschooling neighbor, is begging you for good science, quick! We have suggestions. Please see our *Purchase and Royalty Options* below.

We are grateful for the work you are doing to help shape tomorrow. We are honored that you are making TOPS a part of your teaching effort. Thank you for your good will and kind support.

Sincerely, Ron Marson

Purchase and Royalty Options:

Individual teachers, homeschoolers, libraries:

PURCHASE option: If your colleagues ask to borrow your book, please ask them to read this copyright page, and to contact TOPS for our current catalog so they can purchase their own book. We also have an **online catalog** that you can access at www.topscience.org.

If you are reselling a **used book** to another classroom teacher or homeschooler, please be aware that this still affects us by eliminating a potential book sale. We do not push "newer and better" editions to encourage consumerism. So we ask seller or purchaser (or both!) to acknowledge the ongoing value of this book by sending a contribution to support our continued work. Let your conscience be your guide.

Honor System ROYALTIES: If you wish to make copies from a library, or pass on copies of just a few activities in this book, please calculate their value at 50 cents (25 cents for homeschoolers) per lesson per recipient. Send that amount, or ask the recipient to send that amount, to TOPS. We also gladly accept donations. We know life is busy, but please do follow through on your good intentions promptly. It will only take a few minutes, and you'll know you did the right thing!

Schools and Districts:

You may wish to use this curriculum in several classrooms, in one or more schools. Please observe the following:

PURCHASE option: Order this book in quantities equal to the number of target classrooms. If you order 5 books, for example, then you have unrestricted use of this curriculum in any 5 classrooms per year for the life of your institution. You may order at these quantity discounts:

2-9 copies: 90% of current catalog price + shipping.

10+ copies: 80% of current catalog price + shipping.

ROYALTY option: Purchase 1 book *plus* photocopy or printing rights in quantities equal to the number of designated classrooms. If you pay for 5 Class Licenses, for example, then you have purchased reproduction rights for any 5 classrooms per year for the life of your institution.

1-9 Class Licenses: 70% of current book price per classroom.

10+ Class Licenses: 60% of current book price per classroom.

Workshops and Training Programs:

We are grateful to all of you who spread the word about TOPS. Please limit duplication to only those lessons you will be using, and collect all copies afterward. No take-home copies, please. Copies of copies are prohibited. Ask us for a free shipment of as many current **TOPS Ideas** catalogs as you need to support your efforts. Every catalog contains numerous free sample teaching ideas.

ISBN 0-941008-89-4

CONTENTS

INTRODUCTION

A. A TOPS Model for Effective Science Teaching
C. Getting Ready
D. Gathering Materials
E. Sequencing Task Cards
F. Long Range Objectives
G. Review / Test Questions

TEACHING NOTES

CORE CURRICULUM

1. Transferring Electrons
2. Like / Unlike Charges
3. Circuit Puzzles
4. Conductor or Insulator?
5. Redistributing Charge
6. Polarization
7. Dancing Circles
8. Build A Bulb Holder
9. Build A Cell Holder
10. Build A Switch
11. Ohm's Law (1)
12. Ohm's Law (2)
13. In Series
14. In Parallel
15. Series or Parallel?
16. Double-Throw Switch
17. Two-Way Switches
18. Switching Maze

19. Build a Galvanometer
20. Test Your Galvanometer
21. Which Path?
22. Variable Resistor
23. Adding Resistances
24. Tiny Fuses
25. Electroscope (1)
26. Electroscope (2)

ENRICHMENT CURRICULUM

27. Electrolysis (1)
28. Electrolysis (2)
29. Electrolysis (3)
30. Build a Wet Cell
31. Build a Storage Cell
32. Charge a Capacitor
33. Ammeter (1)
34. Ammeter (2)
35. Generators and Motors
36. Internal Resistance

REPRODUCIBLE STUDENT TASK CARDS

Task Cards 1-36
Supplementary Page: Lettered Millimeter Scales
Graph Paper

A TOPS Model for Effective Science Teaching...

If science were only a set of explanations and a collection of facts, you could teach it with blackboard and chalk. You could assign students to read chapters and answer the questions that followed. Good students would take notes, read the text, turn in assignments, then give you all this information back again on a final exam. Science is traditionally taught in this manner. Everybody learns the same body of information at the same time. Class togetherness is preserved.

But science is more than this.

Science is also process — a dynamic interaction of rational inquiry and creative play. Scientists probe, poke, handle, observe, question, think up theories, test ideas, jump to conclusions, make mistakes, revise, synthesize, communicate, disagree and discover. Students can understand science as process only if they are free to think and act like scientists, in a classroom that recognizes and honors individual differences.

Science is *both* a traditional body of knowledge *and* an individualized process of creative inquiry. Science as process cannot ignore tradition. We stand on the shoulders of those who have gone before. If each generation reinvents the wheel, there is no time to discover the stars. Nor can traditional science continue to evolve and redefine itself without process. Science without this cutting edge of discovery is a static, dead thing.

Here is a teaching model that combines the best of both elements into one integrated whole. It is only a model. Like any scientific theory, it must give way over time to new and better ideas. We challenge you to incorporate this TOPS model into your own teaching practice. Change it and make it better so it works for you.

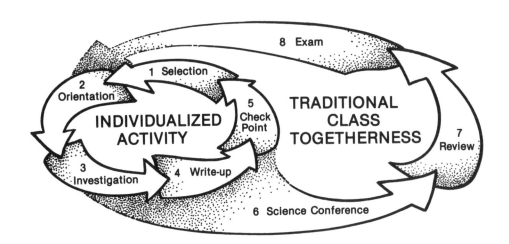

1. SELECTION

Doing TOPS is as easy as selecting the first task card and doing what it says, then the second, then the third, and so on. Working at their own pace, students fall into a natural routine that creates stability and order. They still have questions and problems, to be sure, but students know where they are and where they need to go.

Students generally select task cards in sequence because new concepts build on old ones in a specific order. There are, however, exceptions to this rule: students might *skip* a task that is not challenging; *repeat* a task with doubtful results; *add* a task of their own design to answer original "what would happen if" questions.

2. ORIENTATION

Many students will simply read a task card and immediately understand what to do. Others will require further verbal interpretation. Identify poor readers in your class. When they ask, "What does this mean?" they may be asking in reality, "Will you please read this card aloud?"

With such a diverse range of talent among students, how can you individualize activity and still hope to finish this module as a cohesive group? It's easy. By the time your most advanced students have completed all the task cards, including the enrichment series at the end, your slower students have at least completed the basic core curriculum. This core provides the common background so necessary for meaningful discussion, review and testing on a class basis.

3. INVESTIGATION

Students work through the task cards independently and cooperatively. They follow their own experimental strategies and help each other. You encourage this behavior by helping students only *after* they have tried to help themselves. As a resource person, you work to stay *out* of the center of attention, answering student questions rather than posing teacher questions.

When you need to speak to everyone at once, it is appropriate to interrupt individual task card activity and address the whole class, rather than repeat yourself over and over again. If you plan ahead, you'll find that most interruptions can fit into brief introductory remarks at the beginning of each new period.

4. WRITE-UP

Task cards ask students to explain the "how and why" of things. Write-ups are brief and to the point. Students may accelerate their pace through the task cards by writing these reports out of class.

Students may work alone or in cooperative lab groups. But each one must prepare an original write-up. These must be brought to the teacher for approval as soon as they are completed. Avoid dealing with too many write-ups near the end of the module, by enforcing this simple rule: each write-up must be approved *before* continuing on to the next task card.

5. CHECK POINT

The student and teacher evaluate each write-up together on a pass/no-pass basis. (Thus no time is wasted haggling over grades.) If the student has made reasonable effort consistent with individual ability, the write-up is checked off on a progress chart and included in the student's personal assignment folder or notebook kept on file in class.

Because the student is present when you evaluate, feedback is immediate and effective. A few seconds of this direct student-teacher interaction is surely more effective than 5 minutes worth of margin notes that students may or may not heed. Remember, you don't have to point out every error. Zero in on particulars. If reasonable effort has not been made, direct students to make specific improvements, and see you again for a follow-up check point.

A responsible lab assistant can double the amount of individual attention each student receives. If he or she is mature and respected by your students, have the assistant check the even-numbered write-ups while you check the odd ones. This will balance the work load and insure that all students receive equal treatment.

6. SCIENCE CONFERENCE

After individualized task card activity has ended, this is a time for students to come together, to discuss experimental results, to debate and draw conclusions. Slower students learn about the enrichment activities of faster students. Those who did original investigations, or made unusual discoveries, share this information with their peers, just like scientists at a real conference. This conference is open to films, newspaper articles and community speakers. It is a perfect time to consider the technological and social implications of the topic you are studying.

7. READ AND REVIEW

Does your school have an adopted science textbook? Do parts of your science syllabus still need to be covered? Now is the time to integrate other traditional science resources into your overall program. Your students already share a common background of hands-on lab work. With this shared base of experience, they can now read the text with greater understanding, think and problem-solve more successfully, communicate more effectively.

You might spend just a day on this step or an entire week. Finish with a review of key concepts in preparation for the final exam. Test questions in this module provide an excellent basis for discussion and study.

8. EXAM

Use any combination of the review/test questions, plus questions of your own, to determine how well students have mastered the concepts they've been learning. Those who finish your exam early might begin work on the first activity in the next new TOPS module.

Now that your class has completed a major TOPS learning cycle, it's time to start fresh with a brand new topic. Those who messed up and got behind don't need to stay there. Everyone begins the new topic on an equal footing. This frequent change of pace encourages your students to work hard, to enjoy what they learn, and thereby grow in scientific literacy.

GETTING READY

Here is a checklist of things to think about and preparations to make before your first lesson.

☐ Decide if this TOPS module is the best one to teach next.

TOPS modules are flexible. They can generally be scheduled in any order to meet your own class needs. Some lessons within certain modules, however, do require basic math skills or a knowledge of fundamental laboratory techniques. Review the task cards in this module now if you are not yet familiar with them. Decide whether you should teach any of these other TOPS modules first: *Measuring Length, Graphing, Metric Measure, Weighing* or *Electricity* (before *Magnetism*). It may be that your students already possess these requisite skills or that you can compensate with extra class discussion or special assistance.

☐ Number your task card masters in pencil.

The small number printed in the lower right corner of each task card shows its position within the overall series. If this ordering fits your schedule, copy each number into the blank parentheses directly above it at the top of the card. Be sure to use pencil rather than ink. You may decide to revise, upgrade or rearrange these task cards next time you teach this module. To do this, write your own better ideas on blank 4 x 6 index cards, and renumber them into the task card sequence wherever they fit best. In this manner, your curriculum will adapt and grow as you do.

☐ Copy your task card masters.

You have our permission to reproduce these task cards, for as long as you teach, with only 1 restriction: please limit the distribution of copies you make to the students you personally teach. Encourage other teachers who want to use this module to purchase their *own* copy. This supports TOPS financially, enabling us to continue publishing new TOPS modules for you. For a full list of task card options, please turn to the first task card masters numbered "cards 1-2."

☐ Collect needed materials.

Please see the opposite page.

☐ Organize a way to track completed assignment.

Keep write-ups on file in class. If you lack a vertical file, a box with a brick will serve. File folders or notebooks both make suitable assignment organizers. Students will feel a sense of accomplishment as they see their file folders grow heavy, or their notebooks fill up, with completed assignments. Easy reference and convenient review are assured, since all papers remain in one place.

Ask students to staple a sheet of numbered graph paper to the inside front cover of their file folder or notebook. Use this paper to track each student's progress through the module. Simply initial the corresponding task card number as students turn in each assignment.

☐ Review safety procedures.

Most TOPS experiments are safe even for small children. Certain lessons, however, require heat from a candle flame or Bunsen burner. Others require students to handle sharp objects like scissors, straight pins and razor blades. These task cards should not be attempted by immature students unless they are closely supervised. You might choose instead to turn these experiments into teacher demonstrations.

Unusual hazards are noted in the teaching notes and task cards where appropriate. But the curriculum cannot anticipate irresponsible behavior or negligence. It is ultimately the teacher's responsibility to see that common sense safety rules are followed at all times. Begin with these basic safety rules:

1. Eye Protection: Wear safety goggles when heating liquids or solids to high temperatures.
2. Poisons: Never taste anything unless told to do so.
3. Fire: Keep loose hair or clothing away from flames. Point test tubes which are heating away from your face and your neighbor's.
4. Glass Tubing: Don't force through stoppers. (The teacher should fit glass tubes to stoppers in advance, using a lubricant.)
5. Gas: Light the match first, before turning on the gas.

☐ Communicate your grading expectations.

Whatever your philosophy of grading, your students need to understand the standards you expect and how they will be assessed. Here is a grading scheme that counts individual effort, attitude and overall achievement. We think these 3 components deserve equal weight:

1. Pace (effort): Tally the number of check points you have initialed on the graph paper attached to each student's file folder or science notebook. Low ability students should be able to keep pace with gifted students, since write-ups are evaluated relative to individual performance standards. Students with absences or those who tend to work at a slow pace may (or may not) choose to overcome this disadvantage by assigning themselves more homework out of class.

2. Participation (attitude): This is a subjective grade assigned to reflect each student's attitude and class behavior. Active participators who work to capacity receive high marks. Inactive onlookers, who waste time in class and copy the results of others, receive low marks.

3. Exam (achievement): Task cards point toward generalizations that provide a base for hypothesizing and predicting. A final test over the entire module determines whether students understand relevant theory and can apply it in a predictive way.

Gathering Materials

Listed below is everything you'll need to teach this module. You already have many of these items. The rest are available from your supermarket, drugstore and hardware store. Laboratory supplies may be ordered through a science supply catalog. Hobby stores also carry basic science equipment.

Keep this classification key in mind as you review what's needed:

special in-a-box materials:	general on-the-shelf materials:
Italic type suggests that these materials are unusual. Keep these specialty items in a separate box. After you finish teaching this module, label the box for storage and put it away, ready to use again the next time you teach this module.	Normal type suggests that these materials are common. Keep these basics on shelves or in drawers that are readily accessible to your students. The next TOPS module you teach will likely utilize many of these same materials.
(substituted materials):	***optional materials:**
A parentheses following any item suggests a ready substitute. These alternatives may work just as well as the original, perhaps better. Don't be afraid to improvise, to make do with what you have.	An asterisk sets these items apart. They are nice to have, but you can easily live without them. They are probably not worth the extra trip, unless you are gathering other materials as well.

Everything is listed in order of first use. Start gathering at the top of this list and work down. Ask students to bring recycled items from home. The teaching notes may occasionally suggest additional student activity under the heading "Extensions." Materials for these optional experiments are listed neither here nor in the teaching notes. Read the extension itself to find out what new materials, if any, are required.

Needed quantities depend on how many students you have, how you organize them into activity groups, and how you teach. Decide which of these 3 estimates best applies to you, then adjust quantities up or down as necessary:

$Q_1 / Q_2 / Q_3$

— **Single Student:** Enough for 1 student to do all the experiments.
— **Individualized Approach:** Enough for 30 students informally working in pairs, all self-paced.
— **Traditional Approach:** Enough for 30 students, organized into pairs, all doing the same lesson.

KEY:	*special in-a-box materials* (substituted materials)	general on-the-shelf materials *optional materials

1/1/1	spool thread	2/30/30	baby food jars
2/30/30	*squares of 10 x 10 cm silk cloth*	1/15/15	ceramic magnets, 1 x 3/4 x 1/8 inch — see notes 19
1/5/5	rolls masking tape		
1/15/15	pairs of scissors	1/1/1	box steel wool; fine-grade, unsoaped
2/30/30	styrofoam cups (blocks of polystyrene)	2/30/30	index cards
3/30/30	*size-D dry cells*	1/1/1	*box — a source of corrugated cardboard
10/100/100	*meters bare copper wire, 24 gauge or thinner*	.1/2/2	*liters 5% hydrochloric acid
		1/15/15	*100 mL beakers (baby food jars)
1/6/6	pliers with wire cutting jaws	1/10/15	small test tubes
10/120/130	pennies	.1/2/2	liters saturated baking soda solution
2/30/30	*0.27 amp, 2.33 volt light bulbs - see notes 3*	1/5/5	box matches
2/30/30	straight plastic straws	2/30/30	paper towels
10/200/200	medium or heavy duty rubber bands	1/5/15	*plates or petri dishes
1/1/1	box steel straight pins	1/1/1	bottle 3% hydrogen peroxide dispensed in smaller dropping bottles
1/5/15/	tin cans		
1/15/15	*galvanized nails, about 7 cm or 2 1/2 inches*	6/30/90	*split-shot lead fishing sinkers*
1/2/2	rolls aluminum foil	.1/2/2	liters saturated salt water solution
1/3/6	paper punches	1/5/15	*capacitors rated .1 farad or greater - see teaching notes 32*
5/120/120	clothespins		
1/4/4	boxes paper clips	.1/2/2	cups of oil-based clay
1/15/15	metric rulers	2/10/30	*100 Ω resistors rated at 1/4 or 1/2 watt*
1/15/15	plastic produce bags	1/5/15	*commercial galvanometers — used in activities 32, 35 and 36*
1/4/8	meter sticks		
3/60/60	*meters (or more) insulated 32 gauge wrapping wire — see notes 19*	1/1/1	bottle vinegar
		1/5/15	washers

Sequencing Task Cards

This logic tree shows how all the task cards in this module tie together. In general, students begin at the trunk of the tree and work up through the related branches. As the diagram suggests, the way to upper level activities leads up from lower level activities.

At the teacher's discretion, certain activities can be omitted or sequences changed to meet specific class needs. The only activities that must be completed in sequence are indicated by leaves that open *vertically* into the ones above them. In these cases the lower activity is a prerequisite to the upper.

When possible, students should complete the task cards in the same sequence as numbered. If time is short, however, or certain students need to catch up, you can use the logic tree to identify concept-related *horizontal* activities. Some of these might be omitted since they serve only to reinforce learned concepts rather than introduce new ones.

On the other hand, if students complete all the activities at a certain horizontal concept level, then experience difficulty at the next higher level, you might go back down the logic tree to have students repeat specific key activities for greater reinforcement.

For whatever reason, when you wish to make sequence changes, you'll find this logic tree a valuable reference. Parentheses in the upper right corner of each task card allow you total flexibility. They are left blank so you can pencil in sequence numbers of your own choosing.

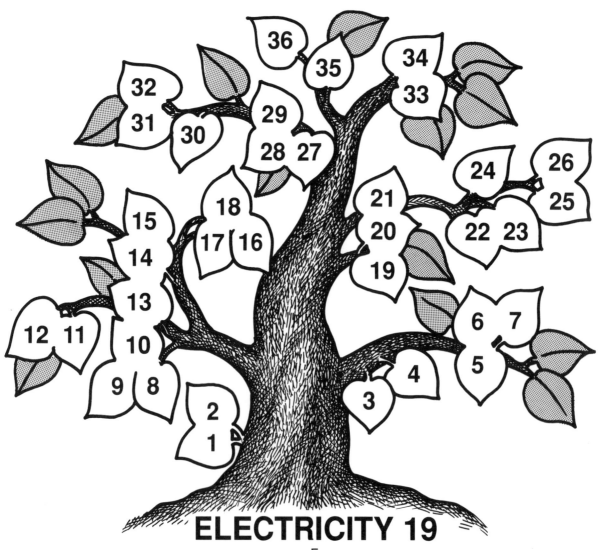

ELECTRICITY 19

E

LONG-RANGE OBJECTIVES

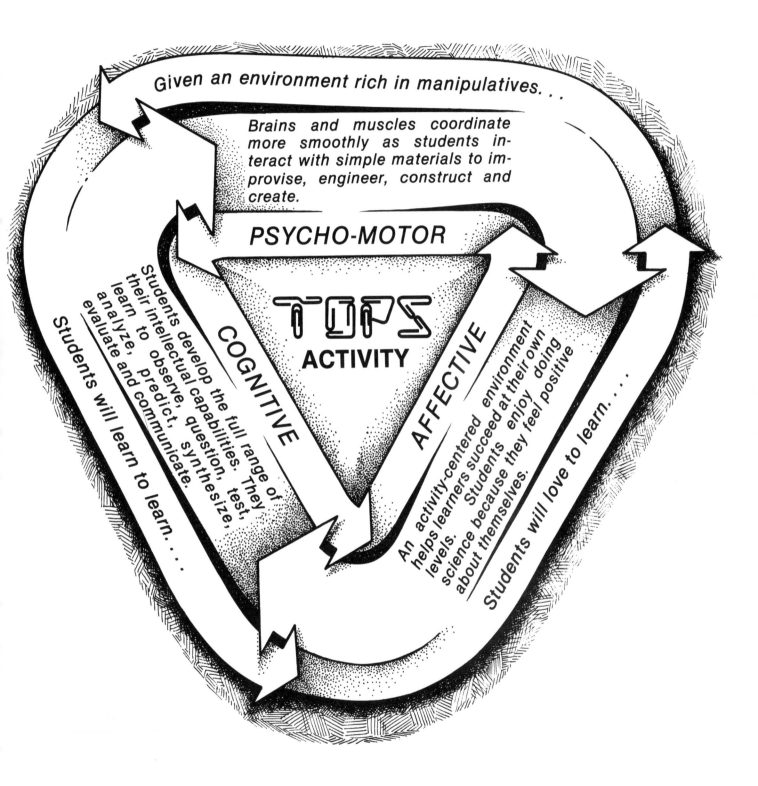

Given an environment rich in manipulatives. . .

Brains and muscles coordinate more smoothly as students interact with simple materials to improvise, engineer, construct and create.

PSYCHO-MOTOR

TOPS ACTIVITY

COGNITIVE

Students develop the full range of their intellectual capabilities. They learn to observe, question, test, analyze, predict, synthesize, evaluate and communicate.

Students will learn to learn. . . .

AFFECTIVE

An activity-centered environment helps learners succeed at their own levels. Students enjoy doing science because they feel positive about themselves.

Students will love to learn. . . .

Review / Test Questions

Photocopy the questions below. On a separate sheet of blank paper, cut and paste those boxes you want to use as test questions. Include questions of your own design, as well. Crowd all these questions onto a single page for students to answer on another paper, or leave space for student responses after each question, as you wish. Duplicate a class set and your custom-made test is ready to use. Use leftover questions as a review in preparation for the final exam.

tasks 1-2
Rubbing a balloon against your shirt sleeve removes electrons from the sleeve and transfers them to the balloon. How is each object now charged, and how will they now interact?

tasks 1-2
a. Why does clothing cling together when taken from an automatic dryer?
b. A scarf removed from the same dryer tends to puff out rather than hang limp. Explain.

tasks 2-3
How does a dry cell manage to push electrons through a wire?

task 3
a. Diagram how to light 1 bulb with 1 dry cell and 1 wire. Use drawings like these:

b. Draw a similar diagram showing how to light 2 bulbs with 1 dry cell and 1 wire.

tasks 3-4
Design an experiment to prove that liquid mercury is a conductor of electricity.

task 4
Is the black ceramic material on a bulb, separating its collar from its end contact point, an insulator or conductor? Explain how you know.

tasks 5-7
A negatively charged balloon is brought close to a tiny ball of neutral aluminum foil resting on a table. Predict how these will interact, giving reasons for your answer.

task 6
A stream of water will bend near a charged comb, but a stream of oil is hardly attracted at all. How might molecules of oil be different from molecules of water?

tasks 8-10
Suppose that you wire together 2 cells, a switch and a bulb into a simple circuit. Then you find that the bulb fails to light when you close the switch.
a. What might be wrong with the bulb holder?
b. What might be wrong with the cells?
c. What might be wrong with the switch? How could you use a single piece of wire to test if the switch was really at fault?

tasks 11-12
How would you use a syringe filled with water connected to a tube to show the relationship between…
a. current and voltage?
b. current and resistance?

task 12
A dry cell operating at 1.3 volts lights a bulb with a resistance of 6.5 ohms. How much current flows through the bulb?

tasks 13-15
Diagram the following circuits. Use appropriate symbols to represent the bulbs, cells, wires and switches.
a. A bulb, dry cell and switch wired in series.
b. Two cells in parallel wired to two bulbs in parallel.
c. Two cells in series wired to three bulbs in parallel, with a switch wired in series to each bulb.

task 14
Why do electricians wire houses in parallel rather than in series?

task 15
Compare each pair of circuits. Decide if the left or the right gives more light.

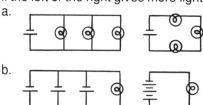

b.

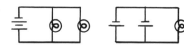

c.

d.

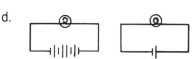

task 16
Diagram a circuit with a double-throw switch that lights just one of two bulbs, but not both at the same time.

task 17
Diagram how you would wire a stairway switch to independently turn the light on or off at both the top and bottom of a stairway.

task 18
Name each switch configuration. Then match each to its most appropriate function defined below.
a.

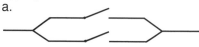

b.

c.

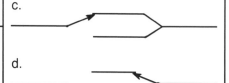

d.

w. Electronically unlocks a high security prison door.
x. Fully controls a yard light from the garage or the house.
y. Smoke detector alarm switches activate a common warning buzzer.
z. Transfers power from a main line to an emergency backup.

task 19
You are given an old dry cell, insulated wire and a compass. Using only these materials, tell how to determine whether the dry cell is dead or alive.

tasks 20, 30
A certain dry cell has terminals with no identifying marks to distinguish positive from negative. How would you use your galvanometer and another normal dry cell to identify the polarity of these terminals?

tasks 21-22
What happens as you flip the switch? Explain.

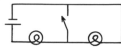

Answers

tasks 1-2
The balloon has an excess of electrons so it is charged negative. The shirt sleeve has a deficiency of electrons so it is charged positive. The balloon will now cling to the sleeve because opposite charges attract.

tasks 1-2
a. As clothing tumbles in an automatic dryer, electrons transfer between the various pieces. This builds up an excess of electrons on some pieces, and creates an electron deficiency on others. Articles with opposite charge are mutually attracted and thus stick together.

b. The scarf has the same excess charge (positive or negative) distributed over its entire surface. Its folds push away from each other, since like charges repel.

tasks 2-3
A dry cell produces electrons at the negative terminal and receives others at the positive terminal. When connected with wire, excess electrons are repelled from the negative end while being attracted to the positive end, thus causing a net migration of electrons through the wire.

task 3
a.   b.

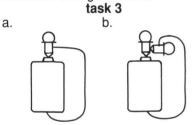

(Other solutions are also possible.)

tasks 3-4
Set a dry cell on end in a beaker containing mercury. Run a wire from this mercury up to the collar of a bulb. Touch the bulb's end to the top of the dry cell like this. The bulb should light, demonstrating that mercury is a conductor of electricity.

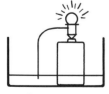

task 4
This black ceramic is an insulator. It prevents electricity from short circuiting between the end contact point and the collar, directing it instead through the filament inside the bulb.

tasks 5-7
Free electrons on the neutral aluminum ball are repelled to its far side as the negative balloon is brought near.

This leaves a positive charge on the near side which, because of its proximity to the balloon, is more strongly attracted than the far side is repelled. This net force lifts the aluminum ball to the balloon. When they touch, electrons flow from the balloon onto the ball, charging it negative, too. This makes it repel off the balloon and back to the table. If these excess electrons then transfer to the table top, the process repeats.

task 6
Molecules of oil don't have charged dipoles (positive charge on one side of the molecule and negative charge on the other). Thus they don't orient themselves to the charged comb, as polar water molecules do, to become attracted as they flow by.

tasks 8-10
a. The bottom of the bulb might not be firmly pressed against its contact wire. The paper clip may have slipped off the rim of the collar. Or the bulb itself may be burned out.

b. The cells may be wired in opposition instead of series. Or they may not be clipped securely together.

c. The contact wires may not touch when the switch is closed. If the switch is really at fault, then the circuit should operate when the faulty switch is by-passed by another piece of wire.

tasks 11-12
a. Push lightly on the syringe (apply a low "voltage") to demonstrate that a small amount of water (or "current") flows from the hose. Then apply more water pressure by pushing the syringe a little harder (turn up the voltage). This shows that increasing voltage creates increasing current.

b. Squeeze the tube to demonstrate how high resistance (a smaller opening) reduces the current. Then relax the tube to show how a low resistance (a larger opening) increases the current in inverse proportion.

task 12
$$I = V/R = 1.3 \text{ volts}/6.5 \text{ ohms}$$
$$= .2 \text{ amperes.}$$

tasks 13-15
a.   b.

c.

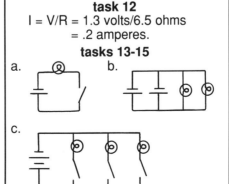

task 14
When connected in parallel, each light or appliance receives full power because it has its own independent path to a common power source. You can turn off the switch to any one or combination, without disrupting service to the others.

task 15
a. left   b. right   c. left   d. right

task 16

task 17

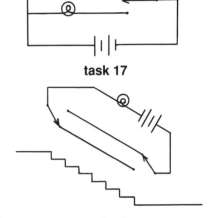

task 18
a. Single switches wired in parallel: y.
b. Single switches wired in series: w.
c. A double throw switch: z.
d. 2-way switches: x.

task 19
Coil the insulated wire around the compass and connect it to the dry cell. If the cell is alive it will send current through the wire, create a magnetic field through the coil and cause the compass needle to deflect. (*For maximum sensitivity the coil should align with the needle, so their magnetic fields will be perpendicular.*)

tasks 20, 30
First establish which way the galvanometer deflects when connected to the terminals with known polarity. Then touch the *same* galvanometer leads to the unknown terminals. If the galvanometer deflects in the same direction, then the known and unknown terminals correspond left to left and right to right. Otherwise they correspond in the opposite direction.

tasks 21-22
When the switch is open, both bulbs light dimly as electrons flow around the perimeter of the circuit. When the switch is closed, the left bulb shines brightly, since it is now on the path of least resistance. The right bulb turns off, since few electrons follow the higher resistance path through both bulbs.

Review / Test Questions (continued)

tasks 21-22

Which lights will short out if you stretch a wire...
- a. across AB?
- b. across CD?
- c. across EF?

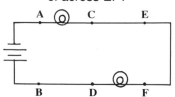

tasks 23, 34

Explain how to connect two or more 10 Ω resistors together to equal each resistance. Show your math.
- a. 20 Ω.
- b. 30 Ω.
- c. 5 Ω.
- d. 15 Ω.
- e. 2 Ω.

task 24

Explain how fuses protect electrical components in a car against damage. Why is it important to use fuses with the correct ampere rating?

task 25

The leaves of an electroscope separate when charged, then slowly come together. Explain why this happens.

task 25

Appliances and tools are often grounded with a third prong. How does this guard against electric shock?

TO GROUND

task 26

The leaves on an electroscope are spread apart. How could you determine whether they are charged positive or negative?

task 26

A pie plate is grounded in the presence of a negatively charged balloon. How is it now charged? Explain.

NEGATIVE CHARGE

task 27

Foil ribbons are connected to 2 dry cells and put into a beaker of weak hydrochloric acid. What gas forms on the right ribbon? The left ribbon? Explain.

task 28

Describe how to capture the gas given off by an Alka-Seltzer tablet dissolving in water.

task 27-29

a. Describe this chemical reaction in words: $HCl \rightarrow H^+ + Cl^-$.
b. Balance this anode reaction:
$Cl^- \rightarrow Cl_2 \uparrow + e^-$.
c. Write a balanced reaction to show the electrolysis of hydrogen ions to hydrogen gas at the cathode.

task 29

Write a balanced chemical equation for the electrolysis of water.

tasks 28-29

Why does an explosion occur when hydrogen and oxygen are ignited? Where did this energy come from?

task 30

A galvanized zinc nail and a copper penny are wired to a galvanometer and placed in a solution of dilute hydrochloric acid. The galvanometer needle deflects, registering current.

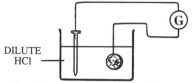

DILUTE HCl

a. How does zinc dissolve off the nail?
b. Why do bubbles form on the penny?
c. Which way do electrons drift through this circuit?

task 30

Using only a penny, an iron washer, and a piece of paper towel soaked in dilute acid, describe how you might make electricity flow through a galvanometer.

task 31

Suppose you want to recharge your car battery. Would you connect it in series or in opposition to your charger? Explain.

task 32

a. A dry cell charges a capacitor in a simple circuit. Which side of the capacitor is positive and which side is negative? Explain.

b. Diagram how to use a double-throw switch so it will charge the capacitor and then discharge the stored energy into a flash bulb.

tasks 33-34

If 20 mA of current deflects the tip of a galvanometer needle through an arc of 3 mm, how far will the same needle be deflected by 100 mA of current?

tasks 33-34

How would you convert a galvanometer into a sensitive ammeter?

tasks 32, 35

Discuss the transformation of energy in each device:
a. A hand cranked generator.
b. A storage battery.
c. An electric motor.
d. A capacitor.
e. A flash bulb.

task 36

How is the distance between a cell's negative and positive terminal related to...
a. its internal resistance?
b. its current output?

Answers (continued)

tasks 21-22
a. Both bulbs will short out.
b. The lower right bulb will short out.
c. Neither bulb will short out.

tasks 23, 34
a. Connect 2 resistors in series:
   $R = 10\ \Omega + 10\ \Omega = 20\ \Omega$.
b. Connect 3 resistors in series:
   $R = 10\ \Omega + 10\ \Omega + 10\ \Omega = 30\ \Omega$.
c. Connect 2 resistors in parallel:
   $1/R = 1/10\ \Omega + 1/10\ \Omega = 1/5\ \Omega$
   $R = 5\ \Omega$.
d. Connect 2 resistors in parallel to another in series:
   $R = 5\ \Omega + 10\ \Omega = 15\ \Omega$.
e. Connect 5 resistors in parallel:
   $1/R = 1/10\ \Omega + 1/10\ \Omega + \ldots = 1/2\ \Omega$
   $R = 2\ \Omega$.

task 24
If current surges in an accidental short circuit, the fuse burns out instead of other more expensive electrical components. This is only possible, however, with fuses of proper ampere ratings — high enough to support normal currents in the line, yet low enough to melt with excess current.

task 25
The leaves separate because they are both holding a like charge and thus repel. They slowly come together again because the electroscope, not perfectly insulated from its surroundings, discharges little by little.

task 25
If a short develops in the appliance, the excess charge will be conducted through this wire into the ground instead of through your body, which has higher resistance.

task 26
Charge a styrofoam cup negative (by rubbing it with a silk cloth), then bring it near the electroscope. If the leaves spread even further apart, then they are charged negative like the cup, becoming increasingly repelled as the cup approaches. If the leaves collapse, then they are charged positive, because the negative cup draws positive charge away from the leaves as it approaches.

task 26
The pie plate is now charged positive. The approaching balloon repelled electrons off the plate into the ground, leaving an excess positive charge behind.

task 27
Hydrogen gas forms on the right as positive hydrogen ions accept electrons from the ribbon. Chlorine gas forms on the left as negative chlorine ions give up electrons to the ribbon.

task 28
Fill a test tube with water. Invert it in a larger beaker of water without letting any air bubbles inside, then place an Alka-Seltzer tablet inside the mouth of the tube. The gas given off by the tablet will bubble up and collect in the bottom of the inverted tube.

task 27-29
a. Hydrochloric acid dissolves (dissociates) in water to form positive hydrogen ions and negative chlorine ions.
b. $2Cl^- \rightarrow Cl_2\uparrow + 2e^-$.
c. $2H^+ + 2e^- \rightarrow H_2\uparrow$

task 29
$$2H_2O \rightarrow 2H_2\uparrow + O_2\uparrow$$

tasks 28-29
Hydrogen and oxygen contain higher energy than water. This energy was given up as an explosion when these gases ignited to form water. It was originally supplied by the dry cells as electrical energy.

task 30
a. Zinc metal dissolves in the acid by forming positive zinc ions and giving up electrons:
   $$Zn \rightarrow Zn^{++} + 2e^-.$$
b. Bubbles form because hydrogen ions in solution accept electrons from the copper penny and form hydrogen gas: $2H^+ + 2e^- \rightarrow H_2\uparrow$
c. Electrons form at the zinc nail, pass through the galvanometer, and are received at the penny.

task 30
Sandwich the soaked piece of paper towel between the penny and the washer, then connect each coin to the galvanometer leads. The iron should dissolve in the acid, forming positive ions and giving up electrons to the circuit. Hydrogen ions in solution should accept electrons from the copper penny, forming hydrogen gas. A sensitive galvanometer should detect electrons passing through this circuit.

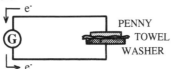

task 31
Connect the car battery to the charger in opposition, so that positive is connected to positive and negative is connected to negative. This will drive electrons backward through the battery in opposition to its normal discharge direction, thus reversing the chemical reactions.

task 32
a.

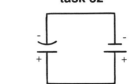

The negative terminal of the dry cell pushes electrons directly into the capacitor where it builds up a negative charge as well. Hence, negative is connected to negative; positive to positive.
b.

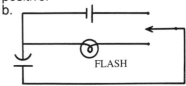

tasks 33-34
A galvanometer needle deflects through a distance that is directly proportional to the current that flows through it. Thus 5 times as much current (100 mA) moves the galvanometer 5 times as far (15 mm).

tasks 33-34
An ammeter is simply a galvanometer that has been calibrated to measure current in amperes. Thus, you need to construct a scale that measures amperes. To do this, measure the amount of needle deflection in the galvanometer for a current of known amperage, then subdivide and multiply this spacing to develop an entire scale.

tasks 32, 35
a. Mechanical energy is converted to electrical energy.
b. Electrical energy is stored as chemical energy.
c. Electrical energy is coverted to mechanical energy.
d. Electrical energy is stored by the capacitor in the same form.
e. Electrical energy is changed into heat and light.

task 36
a. The distance between the terminals is directly proportional to resistance. As this distance increases, so does its internal resistance.
b. The distance is inversely proportional to the cells current output. As this distance increases, its current decreases.

TEACHING NOTES
For Activities 1-36

Task Objective (TO) understand electrostatic attraction between objects as a transfer of electrons.

TRANSFERRING ELECTRONS O   Electricity ( )

1. Tie thread to a 10 x 10 cm patch of silk; tape thread to the bottom of a styrofoam cup.

2. Draw a target circle on the cup. Rub it hard and vigorously with the silk cloth. (This removes electrons from the silk and adds them to the styrofoam.)

3. Hold each object by its thread. Bring one near the other.
     a. Write your observations.
     b. What is the charge on each material? (Remember that electrons are negative.)

4. A charged surface is neutralized (equalized) by touching it or breathing water vapor on it.
     a. Experiment by charging, then neutralizing your silk-styrofoam system. How can you tell this really happens?
     b. How do electrons move so that each charged object is neutralized?
(Save your silk and styrofoam.)

hhhahh...

© 1990 by TOPS Learning Systems   1

Answers / Notes

3a. The silk cloth and styrofoam cup are mutually attracted.
3b. Electrons are removed from the silk, so it is positive. Electrons are added to the styrofoam, so it is negative.

4a. When charged, the cloth and cup cling together. When they are neutralized, there is little or no mutual attraction.
4b. Electrons transfer *from* the negative sytrofoam cup *to* your hands or water vapor. Just the opposite happens with the positive silk cloth: electrons transfer *from* your hands or water vapor to the cloth.

Materials

☐ Thread.
☐ A 10 cm x 10 cm square of silk.
☐ Masking tape.
☐ Scissors.
☐ A cup or other object made from styrofoam (polystyrene).

(TO) observe electrostatic repulsion and attraction between objects of like and unlike charge. To understand why electrons flow through a wire.

LIKE / UNLIKE CHARGES ◯   Electricity (   )

1. Get another styrofoam cup and silk patch to use with your original pair tied to threads.

2. Charge each system, then describe these interactions.

    a. Hold *cup* by thread, then bring *cloth* near.
    b. Hold *cup* by thread, then bring *cup* near.
    c. Hold *cloth* by thread, then bring *cup* near.
    d. Hold *cloth* by thread, then bring *cloth* near.

3. Summarize your findings into a general rule.

4. Examine the labeling on a dry cell.

    a. Which end produces electrons? Which end has a deficit? Explain.
    b. *Predict* what will happen if you connected the ends (terminals) with a wire. Use your results from step 3 to support your answer.
    c. Test your prediction, but for no longer than 5 seconds. (This *short circuit* rapidly drains the cell of energy.) What did you observe?

NEW   ORIGINAL

CHARGE BY RUBBING

© 1990 by TOPS Learning Systems   2

Answers / Notes

2a. The cup and cloth are mutually attracted. They cling together.
2b. The cups are mutually repelled. They stay apart.
2c. The cloth and cup are mutually attracted. They cling together.
2d. The cloths are mutually repelled. They stay apart.

3. Like charges repel. Unlike charges attract.

4. *The term "dry cell", or "cell" for short, is used throughout this module to mean "battery." A battery is technically defined as a collection of 2 or more cells wired together. Since dry cell manufacturers now ignore this distinction and call single cells batteries, it's OK if your students interchange these labels as well.*

4a. The flat end produces an excess of electrons because it is labeled negative. The bump end has a deficiency of electrons because it is labeled positive.

4b. Electrons will flow through the wire. They will be repelled by the negative terminal (like charges repel) and attracted to the positive terminal (unlike charges attract).

4c. The wire quickly heats up as electrons move through it.

Materials

☐ The silk cloth and styrofoam cup, already attached to threads, from the previous activity.
☐ An additional silk cloth and styrofoam cup.
☐ A size-D dry cell. One cell per student is sufficient for all activities in this module if it is used in a responsible manner. To encourage energy conservation, ask students to label their cells by name for their exclusive use, and tell them that dead cells must be replaced at their own expense. Whether you purchase cells from your science budget or ask students to bring them from home, alkaline cells probably provide more power per penny than traditional carbon-zinc cells.
☐ A 30 cm piece of bare copper wire, 24 gauge or thinner. Though length is not important here, 30 cm (the length of a piece of notebook paper) is a convenient length standard for circuit building later on. Copper wire is recommended, though both tin-plated copper wire and aluminum wire may be substituted in all but a few experiments. Choose a thin diameter that is easy to bend and twist; the 22 gauge diameter sold in many variety and hardware stores is too thick.
☐ Wire cutters (optional). The wire can also be bent back and forth until it breaks.

(TO) discover, through trial and error, how to properly wire a circuit. To identify the important contact points on a dry cell and light bulb.

CIRCUIT PUZZLES ○ Electricity ( )

Solve each puzzle using only the specified materials. Diagram each solution with a simple, neat drawing.

BULB   DRY CELL   _ WIRE   PENNY

1. Use only 1 dry cell + 1 wire + 2 pennies: make the wire warm *without* touching it to the cell. (Disconnect as soon as you feel heat to save your dry cell from an energy-draining short circuit.)

2. Use only 1 dry cell + 1 wire + 1 bulb: light the bulb. (Always avoid connections that make the wire warm. The bulb will never light in a short circuit.)

3. Repeat step 2. Diagram at least two more ways to light the bulb. Make each way different.

4. Use 1 dry cell + 1 wire + 1 bulb + 1 penny: light the bulb without touching it to the cell.

5. Work together with a friend. Use 2 dry cells + 1 wire + 1 bulb: make the bulb shine brightly.

6. Work together with a friend. Use 2 dry cells + 1 wire + 2 bulbs: light both bulbs.

7. Identify the important contact points on a dry cell and bulb.

3

Answers / Notes

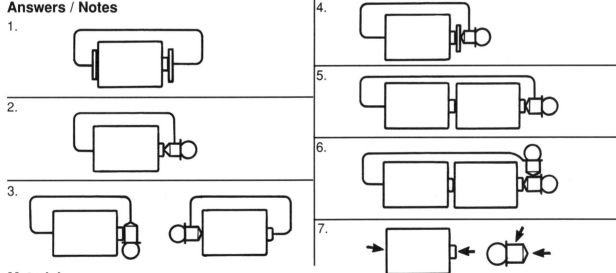

1.

2.

3.

4.

5.

6.

7.

Materials

☐ A size-D dry cell. Students should work in pairs to access the extra cell or bulb.

☐ A 30 cm piece of bare copper wire, 24 gauge or thinner.

☐ Wire cutters (optional).

☐ Two pennies.

☐ A light bulb with collar. Avoid screw-in types. Supply one per student, classified "PR 4," designed for use with two size-D dry cells. Select a high-quality brand rated at or near 2.33 volts and 0.27 amperes. Bulbs with higher amp ratings (perhaps 0.5 amperes, or no stated amp rating) are inferior, drawing perhaps twice the current with no increase in brightness. Radio Shack stores are a good source of bulbs with amp ratings stamped on the collar or explicitly stated on the packaging. Select just one brand for your whole class, with a single set of specifications. Otherwise circuits that students build later, especially parallel constructions, will light unpredictably.

(TO) classify materials as conductors or insulators. To understand how these are used in the design of a light bulb.

CONDUCTOR OR INSULATOR? ○   Electricity ( )

1. Loop a wire 2 times around the neck of a bulb and twist-tie closed. Connect it to a dry cell to make sure it lights.

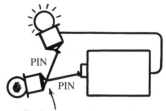

2. Place various objects between the bulb and cell to see if they are conductors (allow electrons to pass) or insulators (don't allow electrons to pass). Make a table that includes these and other objects:

notebook paper	a copper penny	your finger
a nickel	a steel pin	a zinc-galvanized nail
a plastic straw	a tin can	aluminum foil
a rubber band	a wooden pencil	any painted surface…

3. What property do most conductors have in common?

4. Classify these parts on *another* light bulb as either conductors or insulators. (Work with a friend. Use a pin to probe small areas on the bulb.)

5. Diagram how you think the wires inside a bulb are connected. Why do you think ceramic material is placed between the collar and end contact?

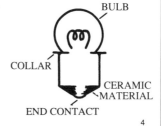

© 1990 by TOPS Learning Systems   4

Answers / Notes

2.

CONDUCTORS		INSULATORS	
nickel	tin can	notebook paper	wooden pencil
copper penny	zinc nail	plastic straw	my finger
steel pin	aluminum foil	rubber band	a painted surface

3. Most conductors are made from metal.

4. The glass bulb and black ceramic material are insulators. The metal collar and end contact point are conductors.
(Students should extend their cell terminals with pins to test the smaller areas. If they experience difficulty, consider substituting 120 V light bulbs.)

End contact is a conductor.

5. The black ceramic material insulates the collar from the end contact, preventing a short circuit that would otherwise bypass the filament in the bulb.

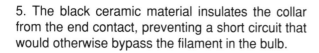

Materials

☐ Copper wire.
☐ Wire cutters (optional).
☐ Light bulbs. (See note 4 above regarding size.)
☐ A dry cell.
☐ All test objects listed in step 2. Most objects you collect here will be used in later activities as well. Select medium-sized galvanized zinc nails about 7 cm (2 1/2 inches) long. These will be turned into chemical cells in activity 30.
☐ A pin.

(TO) induce a redistribution of charge in a neutral conductor. To charge and discharge by contact.

REDISTRIBUTING CHARGE ○   Electricity ( )

1. Cut a 3 cm square of aluminum foil and crumble it into a *loose* ball about as big as a pea. Hang it from your table edge with thread. This ball now has a neutral net charge: equal numbers of positive and negative charge.

2. Charge a styrofoam cup with extra electrons by vigorously rubbing its circle with a silk cloth. Bring the cup *near* (but don't touch) the neutral foil ball.
   a. Write your observations.
   b. Diagram how electrons on the foil ball are *induced* to redistribute.
   c. Explain why the *neutral* ball shows a net *attraction*.

DON'T TOUCH

3. Charge the cup negative once more. This time *touch* it to the foil ball.
   a. Write your observations.
   b. How did the electrons redistribute?

3. TOUCH

4. Touch a wire to the foil ball while it has a negative charge. Does it remain negative? Explain.

4.. TOUCH

5. Do you think a ball of paper will discharge as easily? Write a reasoned prediction, then prove or disprove it by experiment.

5

Introduction

• This ball is electrically *neutral*. It has equal numbers of electrons and protons.

• When a negatively charged comb or a positively charged glass test tube is brought near, electrons on the ball are *induced* to redistribute.

 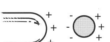

• Why are the ball's electrons induced? (Like charges repel, unlike charges attract.)

• Will the neutral foil ball be attracted? (Yes. Attracting unlike charges are always closer, and therefore stronger, than repelling like charges.)

Answers / Notes

2a. The negative cup attracts the neutral foil ball.

2c. The attracting unlike charges are closer together, and therefore stronger, than the repelling like charges.

2b.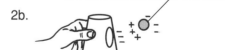

3a. The cup and foil ball are mutually attracted until they touch. Then the ball is repelled away from the cup and they remain apart.

3b. Once the objects touch, some of the excess electrons on the charged cup transfer to the foil ball. This gives both a negative charge, causing them to repel each other.

4. No. After the foil ball is touched by the wire, it is no longer repelled by the negative cup, but attracted as before. Its excess electrons have discharged through the wire, leaving the ball neutral once more.

5. Since paper insulates, students may justifiably reason that it won't accept excess electrons. They'll be surprised that electrons do move onto the paper ball. The wire may also discharge the paper if touched to the right place. But differences are obvious: electrons transfer instantly on the foil ball, sluggishly on the paper ball.

Materials

☐ Aluminum foil.
☐ Thread and masking tape.
☐ The styrofoam cup and silk cloth from Activity 1.
☐ A piece of wire and a metric ruler.

(TO) observe and explain induced polarization in a stream of water and in solid insulators.

POLARIZATION   O   **Electricity ( )**

1a. Charge your styrofoam cup and silk cloth. Bring each one very near (but don't touch) a narrow stream of water. What happens?

1b. Water molecules have equal numbers of protons and electrons (zero net charge), but their atoms are slightly polarized. Use this fact to explain why water is attracted to *both* the cup and the cloth.

$$= O \begin{matrix} \diagdown H^+ \\ \diagup H^+ \end{matrix}$$

2a. Charge the silk and styrofoam again. Bring each one near a neutral glass window and a neutral wood cabinet or door. What happens?

2b. The atoms in glass and wood polarize when an external charge is brought near, shifting the outer electrons slightly off center from the positive nucleus. Use this fact to explain why neutral glass or wood attracts both the positive silk and the negative styrofoam.

6

Answers / Notes

1a. The water stream is attracted to both the negative styrofoam and the positive silk.

1b. The molecules in the water stream are induced to orient themselves relative to the applied charge. If the negative styrofoam is brough near, the molecules turn so that their positive hydrogen side is closer (and thereby more strongly attracted to the styrofoam than the negative oxygen side is repelled). In a similar manner, if the silk cloth is brough near, the molecules turn the opposite way so their negative oxygen side is closer (and thereby more strongly attracted to the silk than the positive hydrogen side is repelled).

2a. The neutral glass and wood attract both the negative styrofoam and the positive silk.

2b. Molecules in the wood and glass are polarized whenever an external charge (positive or negative) is brought nearby. The outer electrons shift slightly *toward* the positive silk, creating temporary polar molecules. This results in a net attraction: the nearer unlike charges attract more strongly than the more widely separated like charges repel. In a similar manner, the outer electrons in wood or glass shift slightly *away* from the negative styrofoam, creating a temporary polarization in the opposite direction. This time the nearer positive side is attracted slightly more than the farther negative side is repelled.

Materials

☐ The styrofoam cup and silk cloth with attached threads.
☐ A source of water.

(TO) polarize insulating and conducting circle punches in an electric field. To compare their charging and discharging characteristics.

DANCING CIRCLES Electricity ( )

1. Use a paper punch to make a group of 6 paper circles and a group of 6 aluminum circles on your desk.

2. Predict how each group of circles will interact with a negatively charged styrofoam cup. Answer each question, stating reasons, *before* you actually do the experiment.

    a. How will the paper circles interact?
    b. How will the foil circles interact?
    c. What differences do you expect to observe between the 2 groups?

3. Test your predictions. Where you at all surprised? Explain.

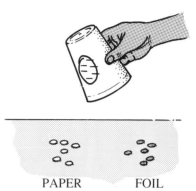

PAPER   FOIL

7

Introduction

Both conductors and insulators polarized in the presence of an external charge, but in different ways.

Because the outer electrons of metal conductors are not bound to any particular atom, they are induced to migrate to opposite sides of the object when an external charge is brought near.

CONDUCTING ATOMS

Because the outer electrons of insulators are tightly held by their respective atoms, the outer electrons polarize locally (shift with respect to their nucleus) but do not migrate great distances.

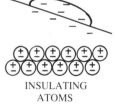

INSULATING ATOMS

Answers / Notes

2a. The neutral paper circles will be attracted. As the negatively charged styrofoam cup is brought near, molecules in the paper will become polarized. The outer electrons will shift away slightly so the nearer positive side will be more strongly attracted than the farther negative side is repelled. The net force thus created will lift the neutral circles from the table to the cup. Once the circles touch the cup, excess electrons may transfer from the cup back into the paper circles, causing them to be repelled back to the table surface.

2b. The neutral foil circles will also be attracted. As the negatively charged styrofoam cup is brought near, the free outer electrons will be repelled to the far side of each circle, thereby concentrating a positive charge on the nearer side of each circle. This will create a net attraction that lifts the neutral circles from the table to the cup. Once the circles touch the cup, excess negative charge will immediately transfer back, causing them to repel back to the table surface.

2c. Electrons will likely exchange more rapidly between the conducting foil punches and cup, than between the insulating paper punches and cup. The foil punches can thus be expected to move more energetically.

3. *Students will not likely predict everything that happens and should report what else they notice.*

    Both kinds of circles jump back and forth as they gain electrons from the cup and lose them again to the table top; but the foil circles "dance" most actively while the paper circles tend to hold their positive charge and stick longer.

Materials

☐ A paper punch.
☐ Aluminum foil.
☐ A styrofoam cup.
☐ A silk cloth.

(TO) construct a bulb holder to use in all circuit building activities.

BUILD A BULB HOLDER ◯   Electricity ( )

1. Follow these directions in alphabetical order.

a. Take the spring off a clothespin. Clamp the 2 halves around a bulb.

b. Wrap about 30 cm of wire 3 smooth turns around a paper clip.

c. Place the end of the paper clip *over* the collar of the bulb. Wrap it tightly in place with a rubber band.

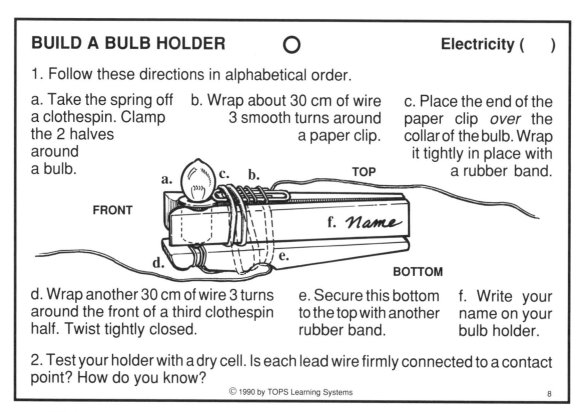

d. Wrap another 30 cm of wire 3 turns around the front of a third clothespin half. Twist tightly closed.

e. Secure this bottom to the top with another rubber band.

f. Write your name on your bulb holder.

2. Test your holder with a dry cell. Is each lead wire firmly connected to a contact point? How do you know?

8

Answers / Notes

1b. *It is important to avoid bunching the wire together when you wrap it 3 turns around the clip. The paper clip must lay flat against both the clothespin and bulb collar after it is wrapped.*

1c. *The paper clip must remain* over *the bulb's collar. This prevents the bulb from sliding up when its contact point presses against the bottom clothespin half in step 1e.*

If your particular bulbs have very narrow collars, it may be necessary to loop the wire above the collar before winding it around the clip.

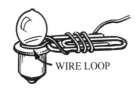

Screw-in bulbs have no collars. It may be possible to screw the bulb directly into the wood so the threads will bite and hold. Otherwise wrap the wire directly around the threads before clamping it between the clothespin halves. After the bulb holder is complete, wrap clear tape over the bulb. Avoid this type of bulb if you can.

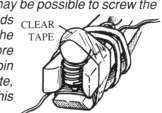

1d. *Again, evenly space these 3 turns for best contact against the bottom of the bulb. Don't bunch the turns together.*

1e. *This rubber band doesn't need to be wound as tightly as the first one. Moderate pressure is sufficient to hold the bottom wire against the bulb without forcing it up excessively against the constraining paper clip.*

2. Yes. Each lead wire is firmly connected to a contact point, because the bulb shines continuously with no flickering.

Materials

☐ Wooden clothespins. Each students will need 3 halves.
☐ A bulb.
☐ Flexible bare copper wire about 24 gauge or thinner; 22 gauge is too thick. Cut these to 30 cm lengths (the length of notebook paper) in advance, or supply wire cutters.
☐ A paper clip.
☐ Rubber bands.
☐ A dry cell.

(TO) construct a cell holder to use in all circuit-building activities.

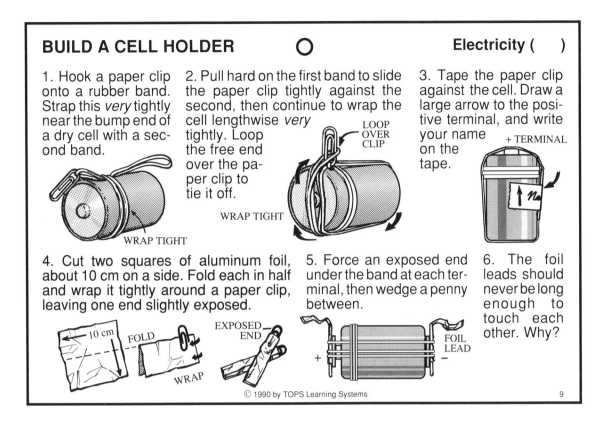

BUILD A CELL HOLDER ⭕ **Electricity ( )**

1. Hook a paper clip onto a rubber band. Strap this *very* tightly near the bump end of a dry cell with a second band.

2. Pull hard on the first band to slide the paper clip tightly against the second, then continue to wrap the cell lengthwise *very* tightly. Loop the free end over the paper clip to tie it off.

LOOP OVER CLIP

WRAP TIGHT

WRAP TIGHT

3. Tape the paper clip against the cell. Draw a large arrow to the positive terminal, and write your name on the tape.

+ TERMINAL

4. Cut two squares of aluminum foil, about 10 cm on a side. Fold each in half and wrap it tightly around a paper clip, leaving one end slightly exposed.

10 cm FOLD EXPOSED END WRAP

5. Force an exposed end under the band at each terminal, then wedge a penny between.

+ FOIL LEAD −

6. The foil leads should never be long enough to touch each other. Why?

© 1990 by TOPS Learning Systems

9

Answers / Notes

5. *Wedging a penny between each foil lead and its terminal insures a good long-term conducting interface that won't deteriorate over time.*

6. If the foil leads are long enough to meet each other, they might accidentally touch and short out the cell.

Materials

☐ Paper clips.
☐ Rubber bands. Thicker rubber bands tend to work better than thinner ones. Any length 4 centimeters or longer will do.
☐ A size-D dry cell. Supply 1 per student.
☐ Masking tape.
☐ Scissors.
☐ Aluminum foil.
☐ A metric ruler.
☐ Pennies.

(TO) construct a switch to use in all circuit building activities. To build a simple circuit.

BUILD A SWITCH

Electricity ( )

1. Replace the spring in a clothespin with a penny. Hold it in place with a rubber band wound to both sides of the coin. Label it with your name.

2. Tightly wind 2 wires that are 30 cm long about 4 turns around each end of the wing tips. The wires should touch only when you press the wings together.

3. Use your pencil to loop the ends of the 2 leads on your switch and the 2 leads on your bulb holder.

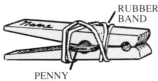
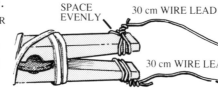
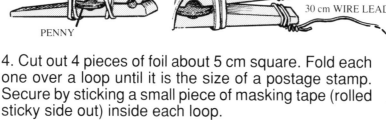

4. Cut out 4 pieces of foil about 5 cm square. Fold each one over a loop until it is the size of a postage stamp. Secure by sticking a small piece of masking tape (rolled sticky side out) inside each loop.

5. Connect your switch, bulb holder and cell holder with paper clips to form a complete circuit. Then close the switch! Describe the path that electrons take as they flow from the negative terminal to the positive terminal.

10

Answers / Notes

1. *Clothespin springs tend to be stiff. The rubber band and penny are substituted for the spring to provide an easy-to-close switch. (Don't wind the rubber band too tightly around the halves or the switch will operate as stiffly as if you had left the spring intact.)*

   If your clothespins are well tapered along each wing, the penny might be left out entirely. Its only function is to provide a sufficient gap across the contact points.

2. *Space out the windings on each wing tip. Don't bunch them together. Thus, if the clothespin doesn't close on center, the windings should still make contact.*

4. *The size and square shape of the original piece of foil is quite arbitrary. Irregular scraps will also fold neatly into postage-stamp sizes. Encourage your class to constructively use foil scraps and conserve resources.*

   Flat foil leads allow circuits to be conveniently paper-clipped together. If you already have a good supply of alligator clips or Fahnstock clips, you may use them instead.

5. Electrons flow from the cell's negative terminal, through the penny, foil, copper wire, paper clip on the bulb holder, bulb collar, bulb filament, bulb's bottom contact, copper wire, across the closed switch, through more copper wire, foil, the penny and back to the positive terminal.

   Students should not conclude that individual electrons zip around the circuit. Rather, a "hornet's nest" of free outer electrons, already present in the wire (or any conductor) drift from negative to positive. This drift speed is remarkably slow, on the order of .5 cm / minute. Even so, the transfer of energy through the wire is nearly instantaneous. Electrons in at the negative terminal immediately push others out at the positive terminal.

Materials

- ☐ A clothespin, penny and rubber band.
- ☐ Copper wire.
- ☐ Wire cutters (optional).
- ☐ The cell and bulb in holders constructed previously.
- ☐ Paper clips.
- ☐ Foil.
- ☐ Scissors and masking tape.
- ☐ A metric ruler.

(TO) develop a kinesthetic sense of current, voltage and resistance. To understand how these variables are interrelated by Ohm's Law.

OHM'S LAW (1)   O   Electricity (   )

Current...	**Voltage...**	**Resistance...**
is the **flow** of electric charge through a circuit. It is measured in **amperes** (A or amps).	is the **force** supplied by a dry cell that causes electric current. It is measured in **volts** (V).	is the **friction** in a wire or bulb that opposes electric current. It is measured in **ohms** (Ω).

1. Squeeze an air-filled bag connected to a straw, and direct the wind against your face.
    a. What models the *current*?
    b. What models the *voltage* that causes this current?
    c. What provides *resistance* to the "voltage" you create?
    d. Squeeze the bag with greater force and less force. How does changing the "voltage" affect the current?
    e. While squeezing the bag with a strong steady "voltage", open and close the straw by pinching it between you fingers. How does changing the resistance affect the current?

2. The German physicist, Georg Ohm (1787-1854), expressed current as a fraction.

$$\text{CURRENT} = \frac{\text{VOLTAGE}}{\text{RESISTANCE}}$$

Explain how this fraction restates what you observed in steps 2d and 2e.

Introduction

Review the 3 variables defined above. Summarize by writing these abbreviated definitions on your blackboard:

CURRENT (**I**)...	VOLTAGE (**V**)...	RESISTANCE (**R**)...
is FLOW measured in AMPERES.	is FORCE measured in VOLTS.	is FRICTION measured in OHMS.

    Write the symbols for each variable on a flash card. Hold up any card asking students to recite the associated definition. Begin with volunteers. Continue with an erased blackboard. End by asking your whole class to chorus back the appropriate response.

Answers / Notes

1a. Current is modeled by the flow of air you feel against your face.
1b. Voltage is modeled by the force you use to squeeze the bag.
1c. Resistance is supplied by the straw. Forcing air through its small opening creates friction.
1d. Increasing the voltage increases the current. Decreasing the voltage decreases the current.
1e. Increasing the resistance decreases the current. Decreasing the resistance increases the current.

2. Current increases or decreases in direct proportion to voltage:

    Current increases or decreases in inverse proportion to resistance:

Materials

☐ A plastic produce bag.
☐ A plastic straw.

(TO) develop a kinesthetic sense of current, voltage and resistance. To practice applying Ohm's Law.

OHM'S LAW (2)   ○   Electricity (   )

1. Ohm's law states that *current (I)* is directly proportional to *voltage (V)* but inversely proportional to *resistance (R)*...

...Explain these ideas in terms of blowing air through your lips.

$$I = \frac{V}{R}$$

2. Current, voltage and resistance are measured in amperes (A or amps), volts (V) and ohms (Ω). Apply Ohm's law to each problem below. (Though air is not actually measured in electrical units, it behaves in a similar way.)

$$1 \text{ ampere} = \frac{1 \text{ volt}}{1 \text{ ohm}}$$

   a. How many "amperes" of air are supplied by blowing with a force of 12 "volts" through your lips held at a resistance of 6 "ohms"? If you double your voltage while holding your resistance constant, how does this change your current?

   b. How many volts are required to force 4 amperes of air through your lips held at a resistance of 5 ohms? If you double your resistance while holding your voltage constant, how does this change your current?

   c. A certain light bulb draws .2 amperes of current when it is connected to a source of 2 volts. What is the resistance of this bulb?

12

Introduction

Use concrete numbers to illustrate how I changes in direction proportion to V (R is held constant):
   5 = 10/2, 10 = 20/2, 15 = 30/2, 20 = 40/2, 25 = 50/2, ..., I = V/R.
In a similar manner, show how I changes in inverse proportion to R (V is held constant):
   60 = 60/1, 30 = 60/2, 20 = 60/3, 15 = 60/4, 12 = 60/5, 10 = 60/6, ..., I = V/R.

Answers / Notes

1. Current (the flow of air through your lips) increases or decreases in direct proportion to voltage (how hard you force air out of your lungs). Blowing harder makes more air flow; blowing softer makes less air flow.

   Current (the flow of air through your lips) increases or decreases in inverse proportion to resistance (how wide you open your lips). Increasing your resistance (closing your lips) makes less air flow; decreasing your resistance (opening your lips) allows more air to flow.

2a. I = V/R = 12 V/6 Ω = 2 A. Doubling the voltage (from 12 V to 24 V) doubles the current (from 2 A to 4 A) when the resistance is held constant (at 6 Ω).

2b. V = I R = 4 A x 5 Ω = 20 V. Doubling the resistance (from 5 Ω to 10 Ω) halves the current (from 4 A to 2 A) when the voltage is held constant (at 20 V).

2c. R = V/I = 2 V/.2 A = 10 Ω.

Materials

None required.

(TO) diagram and build circuits in series. To explain current variations in terms of Ohm's law.

IN SERIES **Electricity ( )**

1. Diagram each circuit:
   a. 1 cell, 1 bulb and 1 switch in series.
   b. 1 cell, 2 bulbs and 1 switch in series.
   c. 2 cells, 1 bulb and 1 switch in series.
   d. 2 cells in opposition connected to 1 bulb and 1 switch in series.

2. Build each circuit as you have diagramed it. Under each drawing, note the relative brightness of the bulbs: bright, medium, dim, none.

3. A bulb's brightness indicates how much current is flowing through it. Under each drawing note the relative amount of current that flows through each circuit you made: high amperes, medium amperes, low amperes, no amperes.

4. In series hook-ups, voltage increases with the number of cells; resistance increases with the number of bulbs.

MORE CELLS, MORE VOLTAGE   MORE BULBS, MORE RESISTANCE

Use this information plus Ohm's law to account for the amount of current in each of your circuits.

13

Introduction

Review how to draw simple circuit diagrams using these symbols.

Put these symbols together to represent a complete circuit:

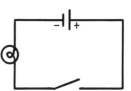

CELL   BULB   SWITCH   WIRE

Demonstrate different ways to connect cells, bulbs and switches in series and opposition.

4 cells   2 cells   4 cells   3 bulbs   2 switches
in series   in opposition   in opposition   in series   in series

Answers / Notes

1-3. a. 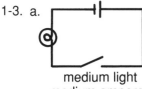   b. 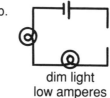   c. 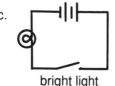   d.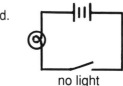

medium light   dim light   bright light   no light
medium amperes   low amperes   high amperes   no amperes

4. Circuit "c" carries the most current because it has the most voltage (2 cells).
   Circuit "a" carries less current than "c" because it has less voltage (1 cell) with the same resistance (1 bulb).
   Circuit "b" carries less current than "a" because it has more resistance (2 bulbs) with the same voltage (1 cell).
   Circuit "d" has no measured current because it has little or no voltage. The 2 cells push against each other to cancel the voltage.

Materials

☐ The basic circuit components constructed in previous activities: 2 bulbs, 2 cell and a switch. Students working in pairs can pool together enough parts to complete all circuits.

(TO) diagram and build a parallel circuit. To understand the advantage of parallel wiring.

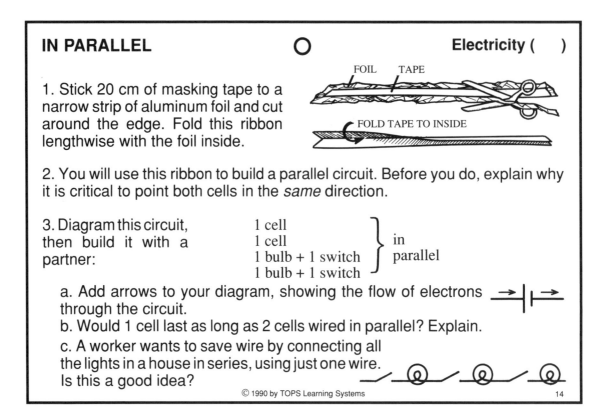

IN PARALLEL ○ Electricity ( )

1. Stick 20 cm of masking tape to a narrow strip of aluminum foil and cut around the edge. Fold this ribbon lengthwise with the foil inside.

FOIL   TAPE

FOLD TAPE TO INSIDE

2. You will use this ribbon to build a parallel circuit. Before you do, explain why it is critical to point both cells in the *same* direction.

3. Diagram this circuit, then build it with a partner:

1 cell	
1 cell	in
1 bulb + 1 switch	parallel
1 bulb + 1 switch	

   a. Add arrows to your diagram, showing the flow of electrons through the circuit.

   b. Would 1 cell last as long as 2 cells wired in parallel? Explain.

   c. A worker wants to save wire by connecting all the lights in a house in series, using just one wire. Is this a good idea?

© 1990 by TOPS Learning Systems   14

Introduction

Demonstrate various ways to connect cells, bulbs and switches in parallel.

3 cells in parallel   4 bulbs in parallel   2 switches in parallel   2 cells and 2 bulbs in parallel

Answers / Notes

1. *These conducting ribbons will be used throughout this module, in all circuits employing parallel construction.*

2. Pointing both cells in the same direction prevents an energy-draining short circuit.

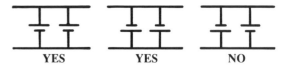

YES   YES   NO

3, 3a.

3b. No. One cell would wear out faster because it would have to supply all the current instead of sharing the load with another cell.

3c. No. In series, the lights cannot be turned on and off independently. One bulb turned off (or burned out) would break the circuit and turn off all the other lights, as well. Parallel wiring may use extra wire, but it overcomes this problem of "all or nothing."

Materials

☐ A metric ruler.
☐ Masking tape, aluminum foil and scissors.
☐ Circuit components: 2 bulbs, 2 cells, 2 switches and 2 conducting ribbons. Students should work in pairs.
☐ Paper clips.

(TO) discover that voltages add as cells are connected in series.

SERIES OR PARALLEL?   Electricity ( )

1. Build the circuit on the left. Rate the bulb's brightness (on a scale of 1 to 5) in the table. Then do the same for the circuit on the right.

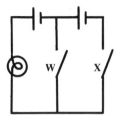

closed switches	bulb brightness
w only	2
x only	
y only	
z only	
y and z	

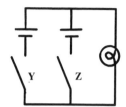

2. Why is it a bad idea to close switch w and x at the same time?

3a. Which arrangement produced more current — 2 cells wired in series or in parallel? Refer to the results in your table to support your answer.
3b. Is this increase caused by more voltage or less resistance? Explain.

4a. Does the voltage increase significantly as you add cells in parallel? Refer to the results in your table to support your answer.
4b. What do you gain by wiring cells in parallel?

15

Answers / Notes

1. *The first number "2" in the table is arbitrary, setting a brightness standard against which other switch combinations can be compared.*

closed switches	bulb brightness
w only	2
x only	5
y only	2
z only	2
y and z	2

2. Closing switch w and x at the same time creates an energy-draining short circuit through the cell that is wired in series between them.

3a. Two cells wired in series (switch x closed) produced light rated at "brightness 5", whereas two cells wired in parallel (switches y and z closed) only produced light rated at "brightness 2". The cells in series, therefore, produced more current.
3b. Resistance in both the parallel and series circuits is contributed mainly by the same bulb, so it is about the same in each circuit. Therefore, the higher current in the series circuit must be caused by an increase in voltage.

4a. Two cells wired in parallel (switch y *and* z closed) produced light rated at "brightness 2". This was about the same brightness as either cell connected alone (switch w closed *or* switch y closed). Adding cells in parallel does not significantly increase current.
4b. Electricity is supplied by all parallel cells in the circuit. Because they share the energy load, they will last longer.

Materials

☐ Circuit components: 1 bulb, 2 cells, 2 switches and 2 conducting ribbons. Students should work in pairs.
☐ Paper clips.

(TO) build a double-throw switch that alternates current through 2 separate bulbs.

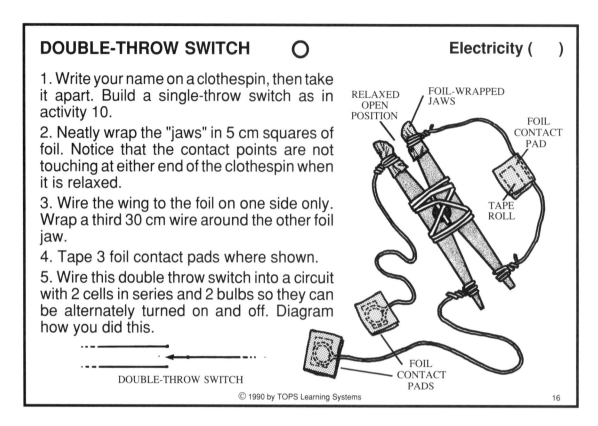

DOUBLE-THROW SWITCH   O   Electricity ( )

1. Write your name on a clothespin, then take it apart. Build a single-throw switch as in activity 10.

2. Neatly wrap the "jaws" in 5 cm squares of foil. Notice that the contact points are not touching at either end of the clothespin when it is relaxed.

3. Wire the wing to the foil on one side only. Wrap a third 30 cm wire around the other foil jaw.

4. Tape 3 foil contact pads where shown.

5. Wire this double throw switch into a circuit with 2 cells in series and 2 bulbs so they can be alternately turned on and off. Diagram how you did this.

DOUBLE-THROW SWITCH

RELAXED OPEN POSITION

FOIL-WRAPPED JAWS

FOIL CONTACT PAD

TAPE ROLL

FOIL CONTACT PADS

16

Introduction

Diagram these switch configurations on your blackboard, identifying each by name.

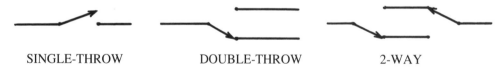

SINGLE-THROW   DOUBLE-THROW   2-WAY

Answers / Notes

5.

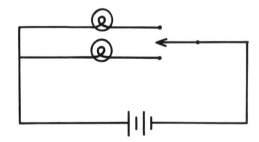

Materials

☐ A clothespin. To conserve materials, assign pairs of students to construct just one switch between them. This applies to the next activity as well.
☐ Copper wire.
☐ Wire cutters (optional).
☐ A penny.
☐ A rubber band.

☐ Scissors.
☐ Aluminum foil.
☐ A metric ruler.
☐ Masking tape.
☐ Circuit components: 2 bulbs, 2 cells and 2 conducting ribbons.
☐ Paper clips.

(TO) build a set of two-way switches that independently operate a bulb.

TWO-WAY SWITCHES Electricity ( )

1. Replace the springs in 2 clothespins with rubber bands.

2. Wrap each jaw, "uppers" and "lowers," in pieces of foil. These jaws should *break* contact when the *wings* are squeezed together.

3. Point the jaws in opposite directions, then join the clothespins with 2 inner wires as shown.

4. Add an outer wire to each clothespin. The *wings* should *make* contact when squeezed together.

5. Attach foil contact pads with rolled tape at the middle of each outer wire. Write your name on your switch.

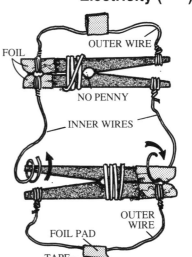

6. Wire your 2-way switch into a series circuit with a bulb and 2 dry cells.
   a. Diagram how your circuit works.
   b. Why are these switches ideal for a stairway?
   c. Your switch stays normally open. How could you modify it to stay closed?

17

Answers/ Notes

1. *Notice that this switch doesn't have a penny in between the clothespin halves. This spacer is unnecessary because the clothespin jaws remain normally closed, not open.*

2-4. *Demonstrate how to twist the clothespin halves into an "x" shape: This makes the ends easier to wrap in foil and wires.*

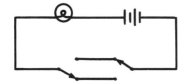

6a.

6b. They make great stairway switches because you can turn a light on at the bottom of the stairs, see your way to the top, then turn the light back off using the other switch located at the top.

6c. Wire the clothespins so their jaws point in the same direction.

Extension
Construct the same circuit as above, substituting commercial wall switches for your clothespin switches. Be careful not to short out your dry cells; it's easy to do!

Materials
☐ Clothespins. To conserve materials, assign students to build switches in pairs.
☐ Copper wire.
☐ Wire cutters (optional).
☐ Rubber bands.
☐ Scissors.
☐ Aluminum foil.
☐ Circuit components: 1 bulb and 2 cells.
☐ Paper clips.

(TO) trace the flow of electricity through a complex maze of switches.

SWITCHING MAZE   O   Electricity (   )

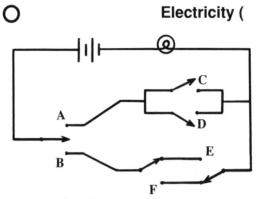

1. Connect 2 single switches in parallel. Connect these to 1 terminal of your double throw switch. Connect a 2-way set of switches to the other terminal. Tie in 2 cells and a bulb like this:

2. List 4 different switching pairs that operate the light.

3. Write a single "and/or" sentence that fully summarizes how to light the bulb.

4. Straighten a paper clip into a single wire. What wires could you interconnect so that...
    a. Switch A alone will operate the bulb?
    b. Switch B alone will operate the bulb?

18

Answers / Notes

2. These switching pairs light the bulb: A-C, A-D, B-F, B-E.

3. To light the bulb, close switches A and C, or A and D, or B and F, or B and E.

4a. Connect opposite sides of switches C and D.
4b. Connect F and E across the center of the 2-way switches. Or connect switch D (on the far side of A) with switch E.

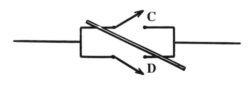

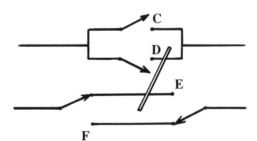

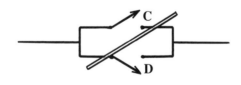

Materials

☐ Two single-throw switches, a double-throw switch, and a 2-way switch.
☐ Circuit components: 1 bulb, 2 cells and 2 conducting ribbons.
☐ Paper clips.

(TO) construct a sensitive galvanometer. To observe that moving electrons create an associated magnetic field.

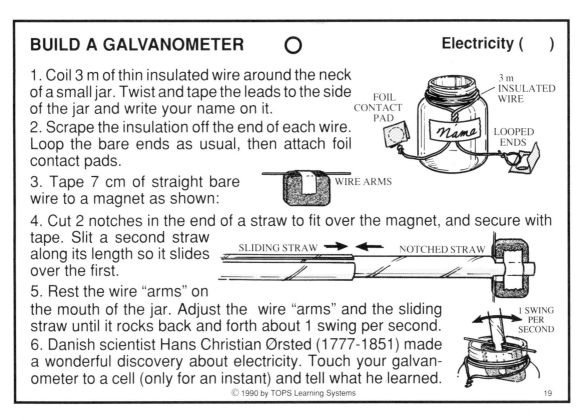

BUILD A GALVANOMETER ◯   **Electricity ( )**

1. Coil 3 m of thin insulated wire around the neck of a small jar. Twist and tape the leads to the side of the jar and write your name on it.

2. Scrape the insulation off the end of each wire. Loop the bare ends as usual, then attach foil contact pads.

3. Tape 7 cm of straight bare wire to a magnet as shown:

4. Cut 2 notches in the end of a straw to fit over the magnet, and secure with tape. Slit a second straw along its length so it slides over the first.

5. Rest the wire "arms" on the mouth of the jar. Adjust the wire "arms" and the sliding straw until it rocks back and forth about 1 swing per second.

6. Danish scientist Hans Christian Ørsted (1777-1851) made a wonderful discovery about electricity. Touch your galvanometer to a cell (only for an instant) and tell what he learned.

© 1990 by TOPS Learning Systems   19

Introduction

Build a galvanometer for your students to inspect. (See task card 33 for a full drawing.) Demonstrate how to fine-tune it for maximum sensitivity without making the straw unstable:

a. Straighten the wire "arms" to stick straight out. The wire should be as free of crooks and bends as possible.

b. Raise the sliding straw until it swings from side to side in about 1 second. (Trim the straws as necessary.)

c. Make the straw stand nearly upright by gently nudging the wire "arms" in the **same** direction as the straw is leaning. (*If it tilts too far from the vertical, this turns the magnet away from its most sensitive position — in perpendicular alignment with the coil's magnetic field.*)

Answers / Notes

1. *The sensitivity of this galvanometer is directly proportional to the number of turns you wind around the neck of the jar. Three meters of wire provide the required threshold sensitivity for all experiments in this module. However, more is better. Doubling the length to 6 meters will double the galvanometer's sensitivity. Doubling several times beyond this will begin to bring your instrument into the sensitivity range of a commercial galvanometer. You may wish to build a more sensitive instrument to demonstrate this.*

6. When electricity moves though a wire it generates an associated magnetic field that will attract a magnet. (*This discovery has transformed science and technology about as much as the invention of the wheel.*)

Materials

☐ A meter stick.
☐ Thin insulated wrapping wire about 30 or 32 gauge. Before you purchase this in an electronics store, survey the discard possibilities in your closet or garage. A burned-out motor, transformer or other piece of junk might contain all the wrapping wire you need and more.
☐ Wire cutters (optional).
☐ A baby food jar or equivalent.
☐ Masking tape.

☐ A knife or sharp pair of scissors to scrape cut away insulation on the wire leads.
☐ Bare copper wire.
☐ A small ceramic magnet about 1 x 3/4 x 1/8 inch. These are commonly sold with or without a center hole at science supply outlets or electronics stores like Radio Shack.
☐ Foil and scissors.
☐ Straight plastic straws.
☐ A dry cell holder.

(TO) observe that galvanometers are sensitive to the direction that electrons flow through a wire. To use this property to determine the relative strengths of two dry cells.

TEST YOUR GALVANOMETER ○   Electricity ( )

RUN CURRENT BOTH WAYS

1. Run electricity through the coil of your galvanometer, first in one direction and then the other. (Touch the leads to your cell only for an *instant*.)

   a. Is your galvanometer sensitive to the direction of the current? Explain.

   b. Your galvanometer has a very low resistance. What would happen if you connected it to your dry cell for an extended period of time? Use Ohm's law to support your answer.

2. Adjust your galvanometer straw for maximum sensitivity: it should rock back and forth about once per second.

   a. See if you can detect current when 2 cells are connected in opposition. If both cells are equally matched, you'll need to watch closely.

   b. How can you decide which cell is stronger?

20

Answers / Notes

1a. Yes. The straw on the galvanometer deflects in opposite directions when you reverse the cell and thus the direction of current through the wire.

1b. Ohm's law predicts that a cell will send current through a closed circuit in inverse proportion to the resistance in the line. If it is connected to a low resistance galvanometer and nothing more, a large current will flow, similar to a short circuit. Connected over an extended period, the cell would be drained of all its energy.

2a. With one strong cell and one weak cell, the galvanometer responds clearly; with 2 cells that are equally balanced, the straw may hardly swing at all. (*To detect a very weak current requires a well adjusted galvanometer used in a room with no moving air. If no movement is immediately apparent, try turning the galvanometer on and off in rhythm with the natural frequency of the straw. If all else fails, substitute a commercial galvanometer or use 1 newer cell and 1 older cell.*)

2b. First connect the cells in opposition and note which way the galvanometer straw deflects. Without changing the orientation of either cell, observe which way the galvanometer straw deflects when it is connected to each cell individually. The cell that deflects the straw in the *same* direction must have been the *stronger*. It determined the direction of current flow when both were connected in opposition. The cell that deflects the straw in the *opposite* direction must have been the *weaker*. It was forced to run in reverse *(and thereby recharged)* when connected in opposition.

Materials

☐ The galvanometer constructed in the previous activity.

☐ Two cells. See note 2a above. Challenge older, more capable students to use their own cells in holders. They will be of nearly equal strength. Provide younger students with a new cell and an old cell secretly marked or coded so they won't know which is which, but you can identify them after the experiment is over.

☐ Paper clips.

(TO) observe how current flows through wire in inverse proportion to its resistance, in accordance with Ohm's law.

WHICH PATH?   Electricity ( )

1. Wire 2 cells, 2 bulbs and 2 switches into a circuit like this:

2. How many bulbs light when you turn on…
   a. Switch x only?
   b. Switch y only?
   c. Both switch x and y?

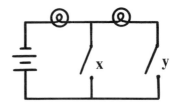

3. Apply Ohm's law to answer each questions:
   a. Why does 1 bulb shine brighter than 2?
   b. Why does only 1 bulb shine when both switches are closed?
   c. Might *some* current travel through the dark bulb when both switches are closed?

4. Test your hypothesis in step 3c: Wire your galvanometer in series, between switch y and its bulb. See if you can detect any current flowing through the dark bulb as you open and close switch y while keeping x closed.

21

Answers / Notes

2a. One bulb lights.

2b. Two bulbs light.

2c. Only one bulb lights.

3a. One bulb shines brighter than two because it has less resistance and thus, by Ohm's law, allows more current to flow.

3b. When both switches are closed, most of the current follows the lower resistance path through 1 bulb rather than the higher resistance path through 2 bulbs.

3c. Ohm's law predicts there will be *less* current flowing through the wire with 2 light bulbs, but still *some*. Current still passes through the dark bulb, even though there is not enough to make it shine.

4. Yes, a current is detectable. *If the galvanometer is adjusted for sensitivity, students will observe a small but distinct deflection in the straw as current is diverted through the dark bulb. (For best results, open and close switch y in time to the natural swinging frequency of the straw.)*

Materials

☐ Circuit components: 2 bulbs, 2 cells, 2 switches and a conducting ribbon.
☐ Paper clips.
☐ An improvised galvanometer.

(TO) control currents in a circuit by operating a variable resistor. To understand that resistance increases as a wire's length increases.

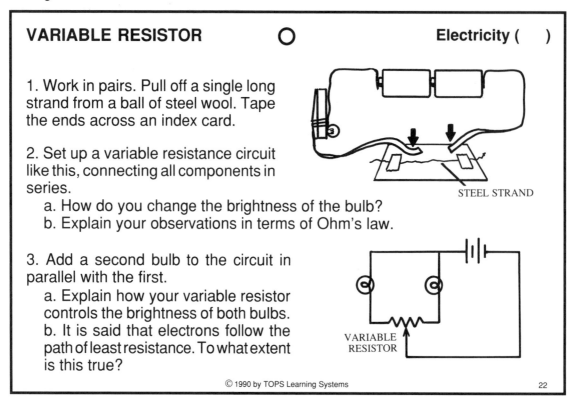

VARIABLE RESISTOR ○ Electricity (   )

1. Work in pairs. Pull off a single long strand from a ball of steel wool. Tape the ends across an index card.

2. Set up a variable resistance circuit like this, connecting all components in series.
   a. How do you change the brightness of the bulb?
   b. Explain your observations in terms of Ohm's law.

3. Add a second bulb to the circuit in parallel with the first.
   a. Explain how your variable resistor controls the brightness of both bulbs.
   b. It is said that electrons follow the path of least resistance. To what extent is this true?

STEEL STRAND

VARIABLE RESISTOR

22

Answers / Notes

2a. To dim the bulb, separate the foil ribbon contact points so the electrons must travel through more of the steel wool fiber. To brighten the bulb, bring these foil contact points closer together.

2b. Ohm's law says that resistance is inversely proportional to current. Increasing the length of the steel wool strand increases resistance in the circuit, thereby reducing the current and dimming the bulb. Likewise, decreasing the length of the steel wool strand decreases resistance in the circuit, thereby increasing the current and brightening the bulb.

3a. Moving the sliding contact nearer either bulb makes it shine brighter while the other shines dimmer. The line with the brighter bulb has a shorter length of steel wool for electrons to travel through, and thus less resistance and more current. The line with the dimmer bulb has a longer length of steel wool for electrons to travel through, and thus more resistance and less current.

3b. This is true to the extent that *more* electrons follow the path of least resistance. Nevertheless *some* electrons do follow the path of greater resistance. Ohm's law predicts that current drops as resistance increases. It does not predict that current stops altogether.

Materials

☐ A fine-grade ball of steel wool without soap. Look for this in hardware stores if supermarkets carry only the soaped variety. Nichrome, a high resistance wire used in light bulbs and heating elements, can be substituted. Remove the coiled filament from a discarded light bulb and stretch it out.
☐ Masking tape.
☐ An index card.
☐ Circuit components: 2 bulbs, 2 cells and 2 conducting ribbons.
☐ Paper clips.

(TO) understand that resistance increases as a wire's diameter decreases. To add resistances in series and in parallel, then observe the combined effect.

ADDING RESISTANCES Electricity ( )

1. Cut 2 index-card squares with sides the length of a paper clip. Then clip as shown.

2. Slide a medium and a thin strand of steel wool under the clips. Tape loose ends underneath.

MEDIUM STRAND   THIN STRAND

TAPE

3. Test *each* resistor with two cells and a bulb. Write your observations and conclusions.

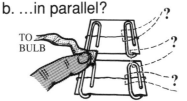

4. What happens to the total resistance when you add both resistors…

a. …in series?

b. …in parallel?

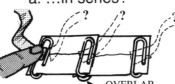

OVERLAP

TO BULB

5. Resistors combine in series and parallel according to these formulas:

$$R_{series} = R_1 + R_2 + R_3 + \ldots \qquad 1/R_{parallel} = 1/R_1 + 1/R_2 + 1/R_3 + \ldots$$

a. If your thinner wire has a resistance of $3\,\Omega$ and your thicker wire $2\,\Omega$, find R_{series} and $R_{parallel}$.

b. Are your experimental results consistent with these formulas? Explain.

23

Introduction

Students without algebra may need some assistance in solving resistance formulas. Work an example on your blackboard as a class exercise. What is the total resistance of three $6\,\Omega$ resistors…

a. connected in series?

$$R_{series} = R_1 + R_2 + R_3$$
$$= 6\,\Omega + 6\,\Omega + 6\,\Omega = 18\,\Omega.$$

b. connected in parallel?

$$1/R_{parallel} = 1/R_1 + 1/R_2 + 1/R_3$$
$$= 1/6\,\Omega + 1/6\,\Omega + 1/6\,\Omega = 3/6\,\Omega = 1/2\,\Omega.$$
Thus $R_{parallel} = 2\,\Omega.$

Answers / Notes

2. *Students should avoid selecting very thick strands of steel wool, as these may not have enough resistance to dim the bulb appreciably.*

3. The thinner strand of steel wool created higher resistance, causing the bulb to shine dimly. The thicker strand of steel wool created lower resistance, causing the bulb to shine brightly. This suggests that wire diameter is inversely proportional to its resistance.

4a. The bulb shined more dimly with both resistors connected in series than with either resistor connected individually. So the total resistance increased.

4b. The bulb shined more brightly with both resistors connected in parallel than with either resistor connected individually. So the total resistance decreased.

5a. $R_{series} = 3\,\Omega + 2\,\Omega = 5\,\Omega$ $\qquad 1/R_{parallel} = 1/3\,\Omega + 1/2\,\Omega = 2/6\,\Omega + 3/6\,\Omega = 5/6\,\Omega$
$$R_{parallel} = 6/5\,\Omega = 1.2\,\Omega$$

5b. Yes. $R_{series} = 5\,\Omega$ was dimmer than either resistance connected individually: $5\,\Omega > 3\,\Omega > 2\Omega$. $R_{parallel} = 1.2\,\Omega$ was brighter than either resistance connected individually: $1.2\,\Omega < 2\,\Omega < 3\Omega$.

Materials

☐ An index card and scissors.
☐ Paper clips.
☐ Masking tape.
☐ Steel wool.
☐ Circuit components: 1 bulb, 2 cells and 2 conducting ribbons.

(TO) understand how fuses work to protect circuits from shorts and overloads.

TINY FUSES ○ Electricity (   )

1. Connect a foil ribbon at each end of a bulb and two cells in series. Attach a paper clip to each free end.

2. Gently pull on a piece of steel wool so iron fibers fall onto white scratch paper. Pass current through these tiny strands between the ends of your paper clips.

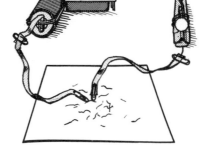

    a. Which fibers tend to melt and burn? Why?
    b. Which fibers tend to sustain the current and light the bulb? Why?

3. A fuse protects bulbs, dry cells, and other circuit components from short circuits. If the current surges, the fuse is the first to melt and break the circuit.

    a. Find a steel wool strand that is just the right size to make a good fuse. How should it perform?

    b. The cells in your circuit push about .25 amperes through the bulb. If you short the bulb, 3 to 15 times more current will surge through the circuit, depending on cell type and age. What size fuse (in amperes) is ideal for this circuit? Defend your answer.

© 1990 by TOPS Learning Systems   24

Answers / Notes

2a. The thin fibers tend to melt and burn because they have higher resistance.

2b. The thicker fibers tend to sustain the current and light the bulb because they have lower resistance.

3. *Traditional carbon-zinc dry cells produce about 0.8A of current in a circuit with no external resistance. Alkaline cells produce many times more than this.*

3a. Students should experiment until they find a fuse of the proper size. They should report that it sustained the lighted bulb, but burned out when the circuit was shorted. Fuses that are too weak won't sustain the lighted bulb; fuses that are too strong won't burn out when the circuit is shorted.

3b. The fuse should be rated somewhat above normal currents expected in the line, but well below the current surge produced by a short circuit. A 1/2 ampere fuse would serve nicely. Rated at twice the normal current expected in the line, it wouldn't burn out accidently with minor current fluctuations, yet it provides adequate short-circuit protection.

Materials

☐ Paper clips.
☐ Circuit components: 1 bulb, 2 cells and 2 foil ribbons.
☐ A ball of steel wool.

notes 24

(TO) build an electroscope and learn to charge it. To explore various ways to discharge it.

ELECTROSCOPE (1)   O   Electricity ( )

1. Punch 2 holes in a piece of aluminum foil. Cut around the holes as close as you can to make 2 leaves like these.

2. Cut a cardboard lid slightly larger than the mouth of the baby food jar. Stick a straight pin through the center, then bend up the point with pliers. Rest your leaves on this hook, and stick the lid on the jar with rolls of tape.

3. Charge a styrofoam cup by rubbing it with silk. Touch it and your finger to the scope.
    a. How does it show that it is charged?
    b. How can you quickly discharge it?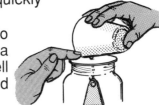

4. Try these ways to discharge (ground) a charged scope. Tell what happens and what you learn.
    a. Touch your finger to the edge of the lid.
    b. Try your pencil point and eraser.
    c. Try a penny on a thread; a can.
    d. Touch the scope with both feet off the ground. (Jump!)
    e. Breathe on the scope as if you were trying to fog a mirror.
    f. Don't do anything. Just observe the leaves for several minutes.

25

Answers / Notes

3a. The leaves separate, indicating they are mutually repelled by excess like charge.

3b. Discharge it by touching the top with your finger. The leaves immediately collapse together.

4a. The electroscope immediately discharges, even when touched near the outer edge of the cardboard. Cardboard has sufficient conducting properties to bleed off the tiny excess charge. (*The electroscope won't discharge, however, if you touch the glass, a better insulator.*)

4b. The pencil discharges the electroscope when touched to its lead, but not when touched to its eraser. Rubber is a good insulator; graphite a good conductor.

4c. The penny on a string discharges the electroscope very slightly; the can on a string a little more. Neither object discharges it completely. This suggests that the electroscope's excess charge redistributes over a wider area, both on the penny or can *and* on the electroscope leaves. More charge is shifted to the can because it has more surface area.

4d. Incredibly, the electroscope discharges only partially when touched while both feet are in the air. (Tapping the scope anywhere on the lid while air-borne will do.) It discharges completely when you connect the scope through your body to the ground. This demonstrates that the earth is a bigger charge sink than your body.

4e. Breathing on the top of the scope slowly bleeds charge off the leaves, causing them to gradually collapse. Water vapor that touches the top of the scope carries away the excess charge. (*This is why static electricity experiments work best with low humidity.*)

4f. The charged leaves will collapse all by themselves, though very slowly. This shows that the electroscope is not perfectly insulated from its surroundings, especially water vapor.

Materials

☐ A paper punch.
☐ Aluminum foil.
☐ Scissors.
☐ A circle of cardboard. An index card will also serve, though not as well.
☐ A baby food jar.

☐ Masking tape.
☐ A straight pin.
☐ A pair of pliers. Needlenose pliers work best.
☐ A styrofoam cup and piece of silk cloth.
☐ Thread.
☐ A penny and a can.

(TO) learn to identify the charge on an electroscope. To charge it by contact and by induction.

ELECTROSCOPE (2) Electricity ( )

1. Diagram these different electroscope conditions using (+) and (-) symbols. Use e⁻ ⟶ to show electron movement.

   a. (-) charged scope:
   b. (+) charged scope.
   c. (-) styrofoam nears a (-) scope.

   d. (+) silk nears a (-) scope.
   e. (-) styrofoam nears a (+) scope.
   f. (+) silk nears a (+) scope.

2. Pull up the pin head on your electroscope just a little, and twist a 30 cm wire around it like this:

30 cm WIRE

3. Charge a styrofoam cup. Touch it to the wire so the leaves stay apart.
   a. Determine the charge on the leaves.
   b. Explain how the electroscope was charged *by contact.*

4. Charge your scope as you did in the previous activity. (Touch both your finger and the cup to the scope together.)
   a. Is your scope charged the same as in step 3?
   b. Propose a theory to explain your observations.
   c. Can you charge your scope *by induction,* without directly touching it with the negative cup? Explain.

© 1990 by TOPS Learning Systems   26

Introduction

The leaves on a charged scope *repel* further apart when approached by a *like* charge.

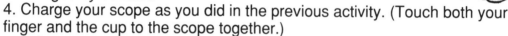

LIKE CHARGES REPEL

But they *attract* closer together when approached by an *unlike* charge.

UNLIKE CHARGES ATTRACT

Answers / Notes

1. a.   b.   c.   d.   e.   f.

2-3. *This wire enables the electroscope to gain a negative charge by direct contact, whereas touching the cup directly to the pin or lid always charges the electroscope positive, by induction. The usual way to charge by induction is to bring the cup near while conducting away repelled electrons onto your finger. Apparently the electrons find a different exit, onto another part of the cup, perhaps.*

3a. The negative styrofoam cup repels the charged leaves even more. Thus they are charged negative.

3b. Excess electrons transferred from the cup through the wire to the leaves. *(It may be possible to hear a tiny spark as electrons jump to the wire.)*

4a. No. The electroscope is now charged positive. The negative cup attracts this positive charge up the pin and away from the collapsing leaves. *(As the cup comes still closer, the leaves collapse altogether into a neutral state. Then they expand again, this time from a build-up of excess negative charge as positive charge continues to flow up the pin toward the approaching cup.)*

4b. Touching the cup *and* your finger directly to the lid repels electrons out of the electroscope onto your finger, leaving the leaves with an overall positive charge.

4c. Yes. Bring the negative cup near (but do not touch) the electroscope while grounding it with your finger. Electrons repel out of the electroscope through your finger, leaving it with an electron deficit or positive charge.

Materials

☐ The electroscope previously constructed, plus a styrofoam cup and piece of silk cloth.
☐ A 30 cm wire. Wire cutters and a ruler are optional.

(TO) observe how hydrogen ions accept electrons to form hydrogen gas, and chlorine ions release electrons to form chlorine gas.

ELECTROLYSIS (1)　　　○　　　　Electricity ( )

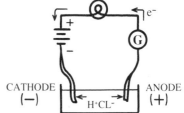

1. Connect 2 dry cells, a bulb and a galvanometer in series with foil ribbons attached to each end.

2. Pour a 5% solution of hydrochloric acid into a jar or beaker, then dip the end of each ribbon in opposite sides of the jar so they don't touch. Write your observations. There is much to notice!

3. Hydrochloric acid forms a solution of positive hydrogen ions (H^+) and negative chlorine ions (Cl^-). These ions receive or give up electrons to form hydrogen gas (H_2) and chlorine gas (Cl_2).

   a. What reaction must be occurring at the negative electrode (the cathode)? Explain.

   b. What reaction must be occurring at the positive electrode (the anode)? Explain.

4. Is electricity best described as a "flow of electrons" or a "flow of charge?" Explain.

5. Diagram how you might collect the gases that bubble off each electrode. (Don't actually do the experiment – chlorine gas is toxic!)

Answers / Notes

2. The galvanometer registers a continuous flow of current through the circuit, though it is not strong enough to light the bulb. Gases are bubbling up from both the positive and negative electrodes. (*Students should not be encouraged to produce significant amounts of gas in this activity, as chlorine is highly toxic.*)

3a. The negative cathode has an excess of electrons, so it must be attracting positive hydrogen ions. These ions must pull electrons off the cathode, since electricity keeps flowing through the circuit. This gain in electrons changes these ions into a neutral gas: $2H^+ + 2e^- \rightarrow H_2 \uparrow$.

3b. The positive anode has a deficit of electrons, so it must be attracting negative chlorine ions. These ions must give up electrons to the anode, since electricity keeps flowing through the circuit. This loss of electrons changes these ions into a neutral gas: $2Cl^- \rightarrow Cl_2 \uparrow + 2e^-$.

4. Electricity is best described as a flow of charge. In this circuit, for example, charge is carried *both* by electrons in the wires and ions in solution.

5. Fill 2 test tubes with the same 5% HCl solution. Cap each one with your thumb, then invert it into a larger beaker or jar filled with the same solution. Push each foil ribbon electrode up into one of the tubes, which trap the gases as they bubble up. (*Students will collect oxygen and hydrogen gas in the next activity using a similar underwater method.*)

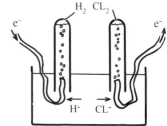

Materials

☐ A 5% solution of hydrochloric acid (add 1 part concentrated HCl to 19 parts water). Although this dilute solution is relatively harmless to clothing, it can still irritate eyes and skin when contacted directly. A label, therefore, should warn students to exercise caution. A saturated solution of salt water will also work in this experiment. Hydrogen and chlorine gases still bubble off each electrode, though the chemical reactions that produce them are more complicated. Be aware, however, that dilute HCl is also the chemical of choice for activity 30.
☐ A beaker or baby food jar.
☐ Circuit components: 1 bulb, 2 cells, 2 conducting ribbons, plus the galvanometer previously constructed.

(TO) separate a test tube of water into oxygen and hydrogen gas.

ELECTROLYSIS (2)　　　　◯　　　　Electricity (　)

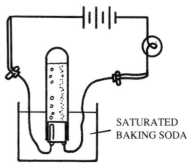

1. Cut a rectangle 2 paper clips long and 1 clip wide from an index card. Roll it the long way around a pencil, so it fits into the mouth of a *small* test tube.

2. Twist tie 30 cm wires to 2 paper clips. Clip these to opposite sides of the rolled card, then push them just into the test tube.

3. Fill the test tube *and* a jar or beaker with a saturated solution of baking soda. Close the test tube with your thumb, and invert it into the jar without letting air inside.

4. Wire this to a bulb plus 3 cells in series. If the bulb lights, your lead wires are shorting out and must be separated.

5. Collect gas until the water level drops to the top of the paper clip electrodes (about 5 minutes). Record your observations.

(Keep the tube inverted in the solution until the next activity.)

28

Answers / Notes

3. *If only a small bubble or two inadvertently gets into the test tube, the experiment should still work.*

4. *Three cells in series force about .3 amperes of current through the bulb, slightly more than it may be designed to take. Even so, with its built in safety tolerance, the bulb will not likely burn out if the electrodes happen to touch briefly.*

5. Many small gas bubbles form on the negative electrode (the cathode). Fewer large gas bubbles form on the positive electrode (the anode).

Extension

One ampere of current delivers 10.44 ml of hydrogen and oxygen per minute. Use this information to calculate how much current flows in your circuit.

Collect gas for exactly 5 minutes, and measure its volume (V) *in milliliters. Use insulated wire to prevent uncollectable gas from forming outside the tube. Set up this proportion to solve for the current* (I) *that runs through your circuit:*

$$\frac{I}{V \text{ mL}} = \frac{1 \text{ A}}{5\,(10.44 \text{ mL})}$$

Materials

☐ An index card.
☐ Scissors.
☐ Paper clips.
☐ A small test tube. This will take 5 to 10 minutes to fill with gas.
☐ Copper wire. A metric ruler and wire cutters are optional.
☐ A small beaker or baby food jar.
☐ A saturated solution of baking soda. Add excess baking soda to water and stir vigorously. Allow to settle, then pour off the clear liquid into a labeled bottle.
☐ Circuit components: a bulb and 3 cells. Students should work in groups of 3 and pool their cells.

(TO) study how current separates water into its component gases. To release stored electrical energy by a chemical reaction.

ELECTROLYSIS (3)   Electricity ( )

1. Electrical energy breaks water apart into 2 gases. What are these gases and which gas forms at each electrode?

2. You can harmlessly burn the gas collected in the inverted tube. To do this:

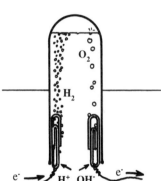

    a. Cover the mouth of the test tube with your finger while it is still submerged.
    b. Continue holding the tube upside-down while a friend strikes a match.
    c. Turn over the test tube, then uncover it directly *under* the match flame.
    d. Write your observations.

3. The gases in the test tube recombined to form water: $2H_2 + O_2 \rightarrow 2H_2O + energy$.

    a. Do hydrogen and oxygen contain more energy as separate gases, or as water?
    b. Energy can neither be created nor destroyed. Where did the energy come from to make your mini-explosion?

© 1990 by TOPS Learning Systems   29

Introduction

This is a great opportunity to expose your students to some basic chemistry:

Four water molecules ($4H_2O$) break into 4 hydrogen ions ($4H^+$) and 4 hydroxide ions ($4OH^-$): $4H_2O \rightarrow 4H^+ + 4OH^-$

H^+ ions are attracted to the negative electrode where they receive electrons and bubble away as hydrogen gas: $4H^+ + 4e^- \rightarrow 2H_2\uparrow$

OH^- ions are attracted to the positive electrode where they give up electrons to create water and oxygen gas bubbles: $4OH^- \rightarrow 2H_2O + O_2\uparrow + 4e^-$

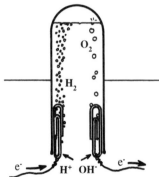

Answers / Notes

1. Water (H_2O) forms from hydrogen and oxygen. Since hydrogen forms positive ions in water, it must be attracted to the negative electrode, where it gains electrons and bubbles away as hydrogen gas. This leaves oxygen to bubble away at the positive electrode. (*Hydroxide ions are attracted to the positive electrode where they give up electrons and combine to form water plus oxygen. See the introduction above.*)

2. *The test tube must be tightly capped while still underwater, and remain tightly covered until a lighted match is placed directly above it. Otherwise the lighter-than-air mixture of oxygen and hydrogen will escape before it can be ignited. This mini-explosion of gas confined to a small test tube is perfectly safe. Exploding more gas in a larger reaction vessel is dangerous and must* NOT *be attempted.*

2d. The gas makes a popping sound: "pheew". There is no discernable heat or flash.

3a. Hydrogen and oxygen contain higher energy as separate gases. They release this energy to form lower-energy water.

3b. Energy to form the gases came from the dry cells. This was released when the gases were ignited.

Materials

☐ The experiment in progress from activity 28.
☐ Matches.

(TO) build a cell that converts chemical energy into electrical energy. To understand this process as the reverse of electrolysis.

BUILD A WET CELL ⭕   Electricity ( )

1. Roll a 3 x 6 cm piece of paper towel around the shaft of a galvanized nail. Leave its head and "neck" exposed.

2. Wind it from the neck down with about 1.5 m of bare copper wire in a single even layer. Wrap in the same direction as the towel is wrapped. *Don't let the wire touch the nail.*

3. Leave about a 6 cm lead at the bottom. Make another lead at the top by wrapping a second wire 2 or 3 times around the metal neck. *This top lead must not touch the wrapped wire.*

4. Set your nail on a plate or petrie dish and soak it with 5% hydrochloric acid. Then drip 4 drops of hydrogen peroxide directly on the nail.

    a. Connect this wet cell to your galvanometer. Can you detect electricity?
    b. In which direction does the current flow? Explain how you know.

5. Acid in the towel dissolves zinc atoms on the galvanized nail, changing them to zinc ions and giving up electrons: $Zn \rightarrow Zn^{++} + 2e^-$. Meanwhile, hydrogen gas bubbles off the copper wire: $2\,H^+ + 2\,e^- \rightarrow H_2\uparrow$.

    a. Diagram how your cell works.   b. Compare this chemical cell to electrolysis.

© 1990 by TOPS Learning Systems   30

Answers / Notes

4a. Yes. The straw deflects when the galvanometer and wet cell are wired together.

4b. Current flows from the zinc nail, through the galvanometer to the coiled copper wire. This can be demonstrated as in activity 20. Substituting a commercial dry cell for the wet cell (positive corresponding to the nail head; negative to the wire coil) causes the galvanometer's straw to deflect in the same direction.

5a.   On surface of zinc nail:

$$Zn \rightarrow Zn^{++} + 2\,e^-$$

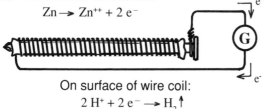

On surface of wire coil:
$$2\,H^+ + 2\,e^- \rightarrow H_2\uparrow$$

5b. Here, chemical energy is changed into electrical energy. This is the reverse of electrolysis, where electrical energy is converted to chemical energy.

Extension

Connect your wet cell to a 1.5 V, .025 A miniature lamp with leads (available from electronic specialty stores like Radio Shack). Let there be light!

Materials

☐ A piece of paper towel.
☐ A medium-sized galvanized nail about 7 cm (2 1/2 inches) long. These are coated with zinc.
☐ Bare copper wire. Do not substitute other kinds of metal wire. A meter stick and wire cutters are optional.
☐ A plate or petri dish. If neither is handy, use plastic wrap or a plastic lid.
☐ A 5% solution of hydrochloric acid. Lemon juice or vinegar may be substituted, though the current output will be less. If a more sensitive commercial galvanometer is substituted for the improvised one, even salt water will dramatically register current.
☐ A 3% solution of hydrogen peroxide (the strength sold as a disinfectant in drug stores). Dispense it in dropping bottles. Hydrogen peroxide prevents cell polarization (a build-up of positive charge on the negative nail) by oxidizing excess hydrogen ions into water. Although this solution will increase the current output, it is not a requirement. The improvised galvanometer will easily detect less current.
☐ An improvised galvanometer.
☐ Paper clips.

(TO) build a working model of a storage cell. To understand why charging and discharging the cell reverses current through the circuit.

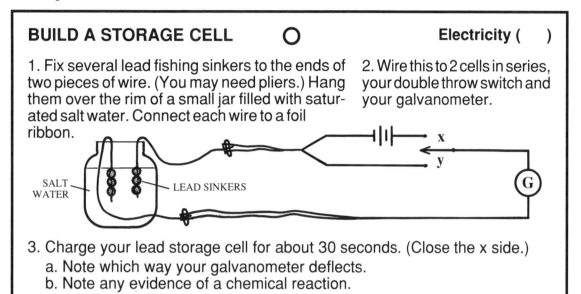

BUILD A STORAGE CELL　　○　　　　　　**Electricity (　)**

1. Fix several lead fishing sinkers to the ends of two pieces of wire. (You may need pliers.) Hang them over the rim of a small jar filled with saturated salt water. Connect each wire to a foil ribbon.

2. Wire this to 2 cells in series, your double throw switch and your galvanometer.

SALT WATER — LEAD SINKERS

x
y
G

3. Charge your lead storage cell for about 30 seconds. (Close the x side.)
　a. Note which way your galvanometer deflects.
　b. Note any evidence of a chemical reaction.
4. Steady the galvanometer needle so it is nearly motionless, then discharge your storage cell through the galvanometer. (Close the y side.)
　a. Does the current still flow in the same direction? Explain.
　b. Summarize how your storage cell converts chemical and electrical energy.

© 1990 by TOPS Learning Systems　　　　　　　　　　31

Answers / Notes

1. *Multiple fishing sinkers on each electrode increase the reacting surface area, and thus the current.*
3a. The galvanometer will deflect right or left depending on how it is connected to the circuit.
3b. Bubbles form at the negative electrode, both on the lead sinkers and the copper wire that come in contact with the salt solution. The positive electrode turns a dull grayish white in contrast to the blacker natural color of lead at the negative electrode. (*These reactions may not happen immediately, but certainly after about 30 seconds. After this initial charging phase, even a 3 second recharge is sufficient to produce an easily recognized discharge through the galvanometer in step 4.*)
4a. No. The galvanometer needle deflects in one direction when you close switch x, and in the opposite direction when you close switch y. This indicates that current flows both ways.
4b. Closing switch x charges the storage cell, converting electrical energy into chemical energy. Closing switch y discharges the storage cell, converting stored chemical energy back into electrical energy.

Materials

☐ Split-shot fishing sinkers made from soft lead. Use as many as necessary to produce a detectable amount of current, probably 2 or 3 on each wire. Pliers may also be needed to squeeze the shot together.
☐ Bare copper wire.
☐ A baby food jar or beaker.
☐ Saturated salt water. Add excess salt to water and stir vigorously. If you prepare this in advance, the opaque white corn stach residue (an additive in table salt) will settle out, allowing you to pour off a clear solution into another jar. This will enable students to have a better view of the chemical reactions occurring at each electrode.
☐ Two dry cells, 2 conducting ribbons and 2 paper clips.
☐ The double-throw switch from activity 16.
☐ A galvanometer.

(TO) charge and discharge a capacitor. To compare its operation to a storage cell.

CHARGE A CAPACITOR ⭕ Electricity ( )

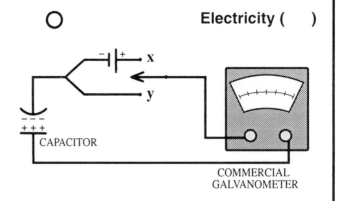

1. Get a capacitor and a commercial galvanometer. Connect these to a dry cell and a 2-way switch like this:

2. Capacitors have plates that store electric charge. Negative charge builds up on one plate while positive charge builds up on the other.

CAPACITOR

COMMERCIAL GALVANOMETER

  a. How do you operate the switch to charge and discharge these plates? How does your galvanometer respond?

  b. Compare this capacitor to the storage cell you made previously.

3. Fully charge your capacitor. Devise a way to confirm that it really does contain a full charge without discharging it.

4. Explain how to give your capacitor a partial charge. How can you tell it's neither "full" nor "empty?"

5. Does your capacitor "leak" charge over time? How do you know?

32

Answers / Notes

2a. Charge the capacitor by closing switch x. Discharge it by closing switch y. The galvanometer needle responds by deflecting in one direction when charging, and in the opposite direction when discharging.

b. The capacitor charges and discharges like the storage cell, causing current to flow both ways through the galvanometer. The capacitor stores energy as a buildup of electric charge between its plates, while the storage cell stores this energy as a chemical reaction between its lead sinkers and salt water.

3. If the galvanometer is fully charged, closing switch x no longer deflects the galvanometer needle because no additional electrons flow to its negative plate. *(The capacitor charges through switch x until its voltage balances the voltage of the dry cell. Once charged, electrons no longer flow.)*

4. To partially charge the capacitor, close switch x only for an instant. It is not fully charged at this point, because closing switch x again still deflects the galvanometer needle. Nor is it fully discharged, because closing switch y also deflects the needle in the opposite direction, but not as much as when the capacitor is fully charged.

5. Yes, the capacitor will lose charge over time, but very slowly: Fully charge the capacitor by closing switch x for several seconds. Release the switch for several more seconds, then close switch x again. The galvanometer needle doesn't deflect because the capacitor retains its full charge over that time period. Then wait perhaps 5 minutes before closing switch x again. This time the needle will deflect a short distance, indicating that the capacitor lost some charge over the longer time interval. *(Wait 24 hours and the needle jumps a little further, but nowhere near the deflection distance of a fully discharged capacitor. If its leads are well insulated, the capacitor retains its charge for a very long time.)*

Materials

☐ A dry cell holder and double-throw switch constructed in previous activities.

☐ A capacitor that can store at least 200 microfarads (MFD) of charge. Capacitors over 1,000 MFD may damage commercial galvanometers. We used a 220 MFD electrolytic capacitor with axial leads purchased at low cost from Radio Shack.

220 µF 35V

☐ A commercial galvanometer. *If you can find a capacitor rated at 0.1 Farad (100,000 MFD) or more, wire this into the circuit with 3 dry cells connected in series. This produces enough current for students to use their improvised straw galvanometer in place of the commercial variety.*

☐ Paper clips, foil and wire.

(TO) quantitatively correlate the deflection of an improvised galvanometer to changes in current.

AMMETER (1)   ○   Electricity (   )

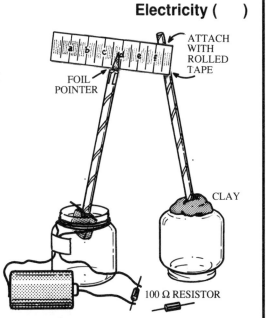

1. Lightly tape a narrow foil pointer to the top of your galvanometer. Adjust the straw to swing from side to side in about 1 second.

2. Put a lump of clay on the top of an inverted baby food jar and stick in a straw. Cut out a paper ruler and fix it to the top with tape rolled sticky side out. Adjust so its millimeter divisions form a scale just behind the freely swinging pointer.

3. Connect the galvanometer to your dry cell, ribbon and a 100 Ω resistor. Keep a second 100 Ω resistor nearby for testing.

4. Work in a space with little or no breeze. Record how many millimeters the pointer deflects when the current passes through 100 Ω. Repeat for 200 Ω, then 50 Ω. Write a full report that includes your data, methods, and conclusions.

© 1990 by TOPS Learning Systems   33

Answers / Notes

4. *This model report is based on a straw that shifts 3 mm with a 100 Ω resistor.*

   100 Ω resistance: The straw indicator was first centered on a major lettered division of the ruler and allowed to come to rest while the circuit was open. Then the circuit was closed and the pointer eventually came to rest at a new position 3 mm to the left (or right) of the original. This shift is reproducible as long as the galvanometer remains undisturbed. *(Shifting the wire arms to a new resting place on the rim of the jar tends to alter the instrument's sensitivity, since these arms are not perfectly straight.)*

   200 Ω resistance: Resistance adds in series. ($R = R_1 + R_2 + ...$) Thus a 200 Ω resistance was made by connecting 100 Ω resistors end to end. (200 Ω = 100 Ω + 100 Ω.) The straw deflected about 1.5 mm when current was run through the galvanometer. As Ohm's law predicts, doubling the resistance cuts the current by half, moving the pointer only half as far.

   50 Ω resistance: Resistance reciprocals add in parallel. ($1/R = 1/R_1 + 1/R_2 + ...$) Thus a 50 Ω resistance was made by connecting both 100 Ω resistors side by side. ($1/R = 1/100 + 1/100 = 2/100 = 1/50$; R = 50 Ω.) This time the straw deflected about 7 mm. As Ohm's law predicts, cutting the resistance in half doubled the current, moving the pointer about twice as far. That the straw deflected 1 mm too far might be attributed to experimental error caused by air currents and uncontrolled variables in the improvised galvanometer.

   A commercial ammeter reveals that current almost (but not quite) doubles as the resistance is divided in two. This is because there is other resistance in the circuit as well — in the wires and in the dry cell itself. Because a cell's resistance increases as it produces more current, this experiment won't produce a simple doubling or halving of the needle's deflection when working with smaller resistors (larger currents).

Materials

☐ Tape, foil and scissors.
☐ A galvanometer.
☐ A lump of oil-based clay.
☐ A baby food jar.
☐ A straw and paper clips.
☐ A cell and conducting ribbon.

☐ A lettered millimeter scale. These are printed 10-up on the back page of this book. Photocopy this image once or twice to serve your entire class.
☐ Two 100 Ω resistors with a 1/4 Watt or 1/2 Watt capacity. These cost perhaps a dime each and are found in electronics stores like Radio Shack.
☐ A work space free of wind currents. If the room is too breezy, only qualitative observations will be possible.

(TO) calibrate an ammeter in milliamps. To use it in measuring current.

AMMETER (2) 　　　O 　　　Electricity (　)

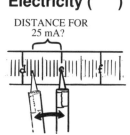
DISTANCE FOR
25 mA?

1. Set up your galvanometer as before, with the millimeter scale resting just behind the pointer. This pointer should move freely, about one swing per second.

2. An average 1.5 volt dry cell connected to a 50 Ω resistance produces about 25 mA of current (.025 A). Through how many mm will this much current deflect the pointer?

3. Knowing how far each 25 mA of current deflects the straw, construct a scale up to 500 mA. Curve it to match the swing of your pointer. (Keep your galvanometer in a safe place so it remains at this setting.)

4. Attach this scale to the straw set in clay as before, so the zero point rests behind your galvanometer pointer. You have just made an ammeter!

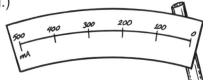

   a. Replace the 50 Ω resistors with your light bulb, and measure the current.
   b. How does the resistance of your bulb compare to a 50 Ω resistance?
   c. Momentarily short out your cell with no light bulb in between. What happens? Why?
   d. What is the difference between a galvanometer and an ammeter?

34

Answers / Notes

2. Answers will vary from about 4 to 8 mm depending on how the galvanometer is adjusted. *(As before, students should center the pointer on a major lettered division of the ruler and allowed it to come to rest while the circuit is open. Then they should connect two 100 Ω resistors in parallel, close the circuit, and note how far the pointer has deflected after again coming to rest. Alert students may ask you why this circuit doesn't have 30 mA of current, in accordance with Ohm's law (1.5 V / 50 Ω = .03 A). Point out that current is reduced by additional resistance in the circuit, especially inside the dry cell.)*

3. *If the galvanometer inadvertently gets bumped out of adjustment (a distinct possibility) advise students to proceed as normal. Before measuring current through a bulb in step 4, they can readjust their galvanometer to conform to the 25 mA spacings on their scales.*

4a. A 1/4 ampere bulb connected to a single dry cell that by now has been well used, may generate 160 mA, give or take 30 mA.

4b. The bulb has 5 or 6 times less resistance. *(Its actual resistance cannot be computed by Ohm's law, since the variable internal resistance of the cell is uncontrolled.)*

4c. The current surges well over 500 mA because there is little resistance in the circuit. *(Students should not wait for the straw to settle down before disconnecting this short circuit.)*

4d. An ammeter is a galvanometer that has been calibrated to read amperes of current.

Extension

Compare your readings from these improvised ammeters to a commercial ammeter. If you have multimeters, remind students to connect ammeters in series (generally across a switch); voltmeters in parallel (generally across a resistance), and ohmmeters outside the circuit (to resistors in isolation).

Materials

☐ Items to improvise an ammeter: A galvanometer with foil pointer, a baby food jar, lump of clay, and straw with lettered millimeter scale.

☐ A cell, conducting ribbon, paper clips plus two 100 Ω resistors.

☐ A work space free from wind currents. This is recommended but not essential.

☐ Tape.

☐ A bulb in its holder.

(TO) model a generator and electric motor. To understand the distinction between AC and DC.

GENERATORS AND MOTORS ○   Electricity (   )

1. Change your galvanometer to a generator:
   a. Remove or fold down the magnet's wire arms. Hook the jar to a commercial galvanometer.
   b. Rest the straw on the jar's rim inside the coil, and rotate it in just one direction.
2. You have just produced electicity:
   a. How did the galvanometer needle move?
   b. What does this tell you about the flow of electricity?
   c. AC means alternating current. DC means direct current. Which kind of electicity does your generator make? Your chemical cell?
3. Generators change mechanical energy to electrical energy, while motors do the opposite.

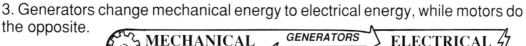 

   MECHANICAL ENERGY ←GENERATORS / MOTORS→ ELECTRICAL ENERGY

   a. Which is the generator in this experiment? Explain.
   b. Which is the motor in this experiment? Explain.

35

Answers / Notes

1. *The improvised galvanometer may be disassembled after this experiment since it won't be used again. It may be convenient to leave the wrapping wire coiled around the rim of the baby food jars. These can be set aside and used the next time you teach this module.*

2a. The galvanometer needle moves back and forth 1 cycle each time you turn the magnet inside the coil through a complete rotation.

2b. It alternates, flowing first in one direction and then in the other direction.

2c. The generator produces AC; the chemical cell produces DC.

3a. The generator is the magnet that you turn inside the coil. Mechanical energy from this spinning magnet changes to electrical energy in the coil around the baby food jar.

3b. The motor is the galvanometer. Electrical energy from the generator travels to a coil inside this galvanometer. Here it creates a magnetic field that mechanically moves a magnet and deflects the indicator needle.

Materials

☐ Parts from the improvised galvanometer.
☐ A commercial galvanometer.
☐ Wire.

(TO) graph a cell's current output as a function of its internal resistance. To verify Ohm's law.

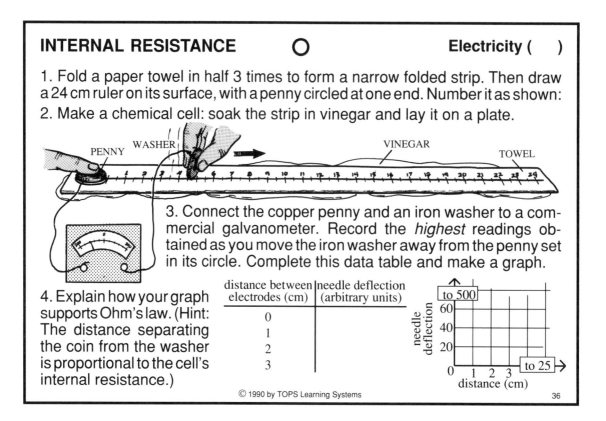

INTERNAL RESISTANCE ○ Electricity ( )

1. Fold a paper towel in half 3 times to form a narrow folded strip. Then draw a 24 cm ruler on its surface, with a penny circled at one end. Number it as shown:
2. Make a chemical cell: soak the strip in vinegar and lay it on a plate.

PENNY   WASHER   VINEGAR   TOWEL

3. Connect the copper penny and an iron washer to a commercial galvanometer. Record the *highest* readings obtained as you move the iron washer away from the penny set in its circle. Complete this data table and make a graph.

4. Explain how your graph supports Ohm's law. (Hint: The distance separating the coin from the washer is proportional to the cell's internal resistance.)

distance between electrodes (cm)	needle deflection (arbitrary units)
0	
1	
2	
3	

36

Introduction

Describe the scale on your commercial galvanometer. (Its center is likely marked zero and numbered 100 to 500 to either side). Count up this scale. (Students should observe that the scale is probably subdivided into units of 20, not 10: 0, 20, 40, 60, 80, 100, 120, etc.)

How is this galvanometer different than an ammeter? (It measures very small quantities of current just like a sensitive ammeter, but its scale is numbered in arbitrary units, not amperes.)

Answers / Notes

3. *The cell tends to polarize over time, causing the voltage to fade. This effect is most pronounced at close electrode distances, causing the galvanometer needle to drop rapidly. For this reason students are instructed to record the highest readings obtained on the galvanometer, before polarization reduces the current.*

4. Resistance (measured as distance) is inversely proportional to current. Qualitatively, this is easy to see. As one variable increases the other variable decreases.

Quantitatively this may be demonstrated by multiplying the resistance and current. The product (an expression of voltage) trends upward with increasing terminal separation, as polarization of the cell becomes less pronounced. Without this polarization, the product remains essentially constant.

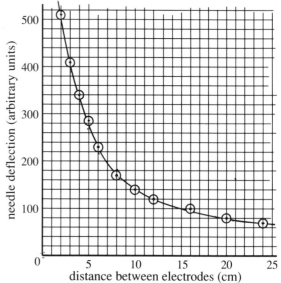

Materials

☐ A soft, absorbent paper towel.
☐ A metric ruler.
☐ A wide plate. Plastic wrap will also serve.
☐ Vinegar, sold in any grocery store. Lemon juice will also work. Do not substitute dilute HCl.

☐ A commercial galvanometer.
☐ A penny, washer and connecting wires.
☐ Graph paper. Photocopy the supplemental grid on the back page of this book.

REPRODUCIBLE
STUDENT
TASK CARDS

Task Cards Options

Here are 3 management options to consider before you photocopy:

1. Consumable Worksheets: Copy 1 complete set of task card pages. Cut out each card and fix it to a separate sheet of boldly lined paper. Duplicate a class set of each worksheet master you have made, 1 per student. Direct students to follow the task card instructions at the top of each page, then respond to questions in the lined space underneath.

2. Nonconsumable Reference Booklets: Copy and collate the 2-up task card pages in sequence. Make perhaps half as many sets as the students who will use them. Staple each set in the upper left corner, both front and back to prevent the outside pages from working loose. Tell students that these task card booklets are for reference only. They should use them as they would any textbook, responding to questions on their own papers, returning them unmarked and in good shape at the end of the module.

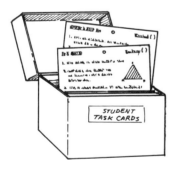

3. Nonconsumable Task Cards: Copy several sets of task card pages. Laminate them, if you wish, for extra durability, then cut out each card to display in your room. You might pin cards to bulletin boards; or punch out the holes and hang them from wall hooks (you can fashion hooks from paper clips and tape these to the wall); or fix cards to cereal boxes with paper fasteners, 4 to a box; or keep cards on designated reference tables. The important thing is to provide enough task card reference points about your classroom to avoid a jam of too many students at any one location. Two or 3 task card sets should accommodate everyone, since different students will use different cards at different times.

TRANSFERRING ELECTRONS ○   Electricity ( )

1. Tie thread to a 10 x 10 cm patch of silk; tape thread to the bottom of a styrofoam cup.

2. Draw a target circle on the cup. Rub it hard and vigorously with the silk cloth. (This removes electrons from the silk and adds them to the styrofoam.)

3. Hold each object by its thread. Bring one near the other.
     a. Write your observations.
     b. What is the charge on each material? (Remember that electrons are negative.)

4. A charged surface is neutralized (equalized) by touching it or breathing water vapor on it.

hhhahh…

     a. Experiment by charging, then neutralizing your silk-styrofoam system. How can you tell this really happens?
     b. How do electrons move so that each charged object is neutralized?

(Save your silk and styrofoam.)

    1

LIKE / UNLIKE CHARGES   ○   Electricity ( )

1. Get another styrofoam cup and silk patch to use with your original pair tied to threads.

NEW   ORIGINAL

2. Charge each system, then describe these interactions.
     a. Hold *cup* by thread, then bring *cloth* near.
     b. Hold *cup* by thread, then bring *cup* near.
     c. Hold *cloth* by thread, then bring *cup* near.
     d. Hold *cloth* by thread, then bring *cloth* near.

CHARGE BY RUBBING

3. Summarize your findings into a general rule.

4. Examine the labeling on a dry cell.
     a. Which end produces electrons? Which end has a deficit? Explain.
     b. *Predict* what will happen if you connected the ends (terminals) with a wire. Use your results from step 3 to support your answer.
     c. Test your prediction, but for no longer than 5 seconds. (This *short circuit* rapidly drains the cell of energy.) What did you observe?

    2

CIRCUIT PUZZLES ◯

Electricity (   )

Solve each puzzle using only the specified materials. Diagram each solution with a simple, neat drawing.

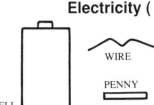

BULB

DRY CELL

WIRE

PENNY

1. Use only 1 dry cell + 1 wire + 2 pennies: make the wire warm *without* touching it to the cell. (Disconnect as soon as you feel heat to save your dry cell from an energy-draining short circuit.)

2. Use only 1 dry cell + 1 wire + 1 bulb: light the bulb. (Always avoid connections that make the wire warm. The bulb will never light in a short circuit.)

3. Repeat step 2. Diagram at least two more ways to light the bulb. Make each way different.

4. Use 1 dry cell + 1 wire + 1 bulb + 1 penny: light the bulb without touching it to the cell.

5. Work together with a friend. Use 2 dry cells + 1 wire + 1 bulb: make the bulb shine brightly.

6. Work together with a friend. Use 2 dry cells + 1 wire + 2 bulbs: light both bulbs.

7. Identify the important contact points on a dry cell and bulb.

© 1990 by TOPS Learning Systems

3

CONDUCTOR OR INSULATOR? ◯

Electricity (   )

1. Loop a wire 2 times around the neck of a bulb and twist-tie closed. Connect it to a dry cell to make sure it lights.

TEST OBJECT

2. Place various objects between the bulb and cell to see if they are conductors (allow electrons to pass) or insulators (don't allow electrons to pass). Make a table that includes these and other objects:

notebook paper	a copper penny	your finger
a nickel	a steel pin	a zinc-galvanized nail
a plastic straw	a tin can	aluminum foil
a rubber band	a wooden pencil	any painted surface…

3. What property do most conductors have in common?

4. Classify these parts on *another* light bulb as either conductors or insulators. (Work with a friend. Use a pin to probe small areas on the bulb.)

5. Diagram how you think the wires inside a bulb are connected. Why do you think ceramic material is placed between the collar and end contact?

BULB

COLLAR

CERAMIC MATERIAL

END CONTACT

© 1990 by TOPS Learning Systems

4

---

cards 3-4

REDISTRIBUTING CHARGE ○ Electricity ( )

1. Cut a 3 cm square of aluminum foil and crumble it into a *loose* ball about as big as a pea. Hang it from your table edge with thread. This ball now has a neutral net charge: equal numbers of positive and negative charge.

2. Charge a styrofoam cup with extra electrons by vigorously rubbing its circle with a silk cloth. Bring the cup *near* (but don't touch) the neutral foil ball.

   a. Write your observations.
   b. Diagram how electrons on the foil ball are *induced* to redistribute.
   c. Explain why the *neutral* ball shows a net *attraction*.

3. Charge the cup negative once more. This time *touch* it to the foil ball.
   a. Write your observations.
   b. How did the electrons redistribute?

4. Touch a wire to the foil ball while it has a negative charge. Does it remain negative? Explain.

5. Do you think a ball of paper will discharge as easily? Write a reasoned prediction, then prove or disprove it by experiment.

© 1990 by TOPS Learning Systems 5

POLARIZATION ○ Electricity ( )

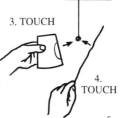

1a. Charge your styrofoam cup and silk cloth. Bring each one very near (but don't touch) a narrow stream of water. What happens?

1b. Water molecules have equal numbers of protons and electrons (zero net charge), but their atoms are slightly polarized. Use this fact to explain why water is attracted to *both* the cup and the cloth.

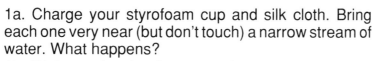

$$= O \begin{smallmatrix} \nearrow H^+ \\ \searrow H^+ \end{smallmatrix}$$

2a. Charge the silk and styrofoam again. Bring each one near a neutral glass window and a neutral wood cabinet or door. What happens?

2b. The atoms in glass and wood polarize when an external charge is brought near, shifting the outer electrons slightly off center from the positive nucleus. Use this fact to explain why neutral glass or wood attracts both the positive silk and the negative styrofoam.

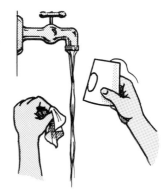

© 1990 by TOPS Learning Systems 6

DANCING CIRCLES **Electricity (   )**

1. Use a paper punch to make a group of 6 paper circles and a group of 6 aluminum circles on your desk.

2. Predict how each group of circles will interact with a negatively charged styrofoam cup. Answer each question, stating reasons, *before* you actually do the experiment.

    a. How will the paper circles interact?
    b. How will the foil circles interact?
    c. What differences do you expect to observe between the 2 groups?

3. Test your predictions. Where you at all surprised? Explain.

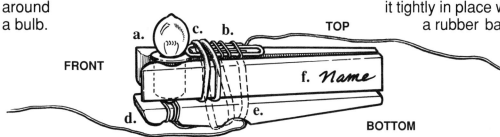

PAPER   FOIL

  7

BUILD A BULB HOLDER **Electricity (   )**

1. Follow these directions in alphabetical order.

a. Take the spring off a clothespin. Clamp the 2 halves around a bulb.

b. Wrap about 30 cm of wire 3 smooth turns around a paper clip.

c. Place the end of the paper clip *over* the collar of the bulb. Wrap it tightly in place with a rubber band.

a. c. b.   TOP

FRONT

f. *Name*

d.   e.

BOTTOM

d. Wrap another 30 cm of wire 3 turns around the front of a third clothespin half. Twist tightly closed.

e. Secure this bottom to the top with another rubber band.

f. Write your name on your bulb holder.

2. Test your holder with a dry cell. Is each lead wire firmly connected to a contact point? How do you know?

  8

BUILD A CELL HOLDER ◯ Electricity (   )

1. Hook a paper clip onto a rubber band. Strap this *very* tightly near the bump end of a dry cell with a second band.

WRAP TIGHT

2. Pull hard on the first band to slide the paper clip tightly against the second, then continue to wrap the cell lengthwise *very* tightly. Loop the free end over the paper clip to tie it off.

LOOP OVER CLIP

WRAP TIGHT

3. Tape the paper clip against the cell. Draw a large arrow to the positive terminal, and write your name on the tape.

+ TERMINAL

4. Cut two squares of aluminum foil, about 10 cm on a side. Fold each in half and wrap it tightly around a paper clip, leaving one end slightly exposed.

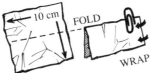

10 cm FOLD WRAP

EXPOSED END

5. Force an exposed end under the band at each terminal, then wedge a penny between.

+   −   FOIL LEAD

6. The foil leads should never be long enough to touch each other. Why?

© 1990 by TOPS Learning Systems

9

BUILD A SWITCH ◯ Electricity (   )

1. Replace the spring in a clothespin with a penny. Hold it in place with a rubber band wound to both sides of the coin. Label it with your name.

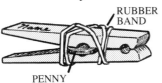

RUBBER BAND

PENNY

2. Tightly wind 2 wires that are 30 cm long about 4 turns around each end of the wing tips. The wires should touch only when you press the wings together.

SPACE EVENLY   30 cm WIRE LEAD

30 cm WIRE LEAD

3. Use your pencil to loop the ends of the 2 leads on your switch and the 2 leads on your bulb holder.

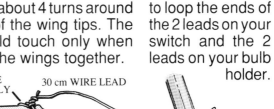

BULB HOLDER

ROLLED TAPE

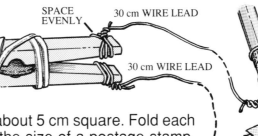

FOIL

4. Cut out 4 pieces of foil about 5 cm square. Fold each one over a loop until it is the size of a postage stamp. Secure by sticking a small piece of masking tape (rolled sticky side out) inside each loop.

5. Connect your switch, bulb holder and cell holder with paper clips to form a complete circuit. Then close the switch! Describe the path that electrons take as they flow from the negative terminal to the positive terminal.

© 1990 by TOPS Learning Systems

10

OHM'S LAW (1)   Electricity (   )

Current...
is the **flow** of electric charge through a circuit. It is measured in **amperes** (A or amps).

Voltage...
is the **force** supplied by a dry cell that causes electric current. It is measured in **volts** (V).

Resistance...
is the **friction** in a wire or bulb that opposes electric current. It is measured in **ohms** (Ω).

1. Squeeze an air-filled bag connected to a straw, and direct the wind against your face.

   a. What models the *current*?

   b. What models the *voltage* that causes this current?

   c. What provides *resistance* to the "voltage" you create?

   d. Squeeze the bag with greater force and less force. How does changing the "voltage" affect the current?

   e. While squeezing the bag with a strong steady "voltage", open and close the straw by pinching it between you fingers. How does changing the resistance affect the current?

2. The German physicist, Georg Ohm (1787-1854), expressed current as a fraction. Explain how this fraction restates what you observed in steps 2d and 2e.

$$\textbf{CURRENT} = \frac{\textbf{VOLTAGE}}{\textbf{RESISTANCE}}$$

11

OHM'S LAW (2)   Electricity (   )

1. Ohm's law states that *current (I)* is directly proportional to *voltage (V)* but inversely proportional to *resistance (R)*...

$$I = \frac{V}{R}$$

...Explain these ideas in terms of blowing air through your lips.

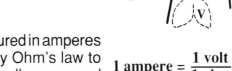

2. Current, voltage and resistance are measured in amperes (A or amps), volts (V) and ohms (Ω). Apply Ohm's law to each problem below. (Though air is not actually measured in electrical units, it behaves in a similar way.)

$$1 \text{ ampere} = \frac{1 \text{ volt}}{1 \text{ ohm}}$$

   a. How many "amperes" of air are supplied by blowing with a force of 12 "volts" through your lips held at a resistance of 6 "ohms"? If you double your voltage while holding your resistance constant, how does this change your current?

   b. How many volts are required to force 4 amperes of air through your lips held at a resistance of 5 ohms? If you double your resistance while holding your voltage constant, how does this change your current?

   c. A certain light bulb draws .2 amperes of current when it is connected to a source of 2 volts. What is the resistance of this bulb?

12

IN SERIES

Electricity ( )

1. Diagram each circuit:
   a. 1 cell, 1 bulb and 1 switch in series.
   b. 1 cell, 2 bulbs and 1 switch in series.
   c. 2 cells, 1 bulb and 1 switch in series.
   d. 2 cells in opposition connected to 1 bulb and 1 switch in series.

2. Build each circuit as you have diagramed it. Under each drawing, note the relative brightness of the bulbs: bright, medium, dim, none.

3. A bulb's brightness indicates how much current is flowing through it. Under each drawing note the relative amount of current that flows through each circuit you made: high amperes, medium amperes, low amperes, no amperes.

4. In series hook-ups, voltage increases with the number of cells; resistance increases with the number of bulbs.

MORE CELLS, MORE VOLTAGE   MORE BULBS, MORE RESISTANCE

Use this information plus Ohm's law to account for the amount of current in each of your circuits.

13

IN PARALLEL
Electricity ( )

FOIL   TAPE

1. Stick 20 cm of masking tape to a narrow strip of aluminum foil and cut around the edge. Fold this ribbon lengthwise with the foil inside.

FOLD TAPE TO INSIDE

2. You will use this ribbon to build a parallel circuit. Before you do, explain why it is critical to point both cells in the *same* direction.

3. Diagram this circuit, then build it with a partner:

1 cell	
1 cell	in
1 bulb + 1 switch	parallel
1 bulb + 1 switch	

a. Add arrows to your diagram, showing the flow of electrons  through the circuit.

b. Would 1 cell last as long as 2 cells wired in parallel? Explain.

c. A worker wants to save wire by connecting all the lights in a house in series, using just one wire. Is this a good idea?

14

---

cards 13-14

SERIES OR PARALLEL?   Electricity ( )

1. Build the circuit on the left. Rate the bulb's brightness (on a scale of 1 to 5) in the table. Then do the same for the circuit on the right.

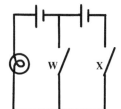

closed switches	bulb brightness
w only	2
x only	
y only	
z only	
y and z	

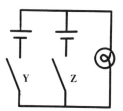

2. Why is it a bad idea to close switch w and x at the same time?

3a. Which arrangement produced more current — 2 cells wired in series or in parallel? Refer to the results in your table to support your answer.
3b. Is this increase caused by more voltage or less resistance? Explain.

4a. Does the voltage increase significantly as you add cells in parallel? Refer to the results in your table to support your answer.
4b. What do you gain by wiring cells in parallel?

15

DOUBLE-THROW SWITCH   O   Electricity ( )

1. Write your name on a clothespin, then take it apart. Build a single-throw switch as in activity 10.

2. Neatly wrap the "jaws" in 5 cm squares of foil. Notice that the contact points are not touching at either end of the clothespin when it is relaxed.

3. Wire the wing to the foil on one side only. Wrap a third 30 cm wire around the other foil jaw.

4. Tape 3 foil contact pads where shown.

5. Wire this double throw switch into a circuit with 2 cells in series and 2 bulbs so they can be alternately turned on and off. Diagram how you did this.

DOUBLE-THROW SWITCH

RELAXED OPEN POSITION

FOIL-WRAPPED JAWS

FOIL CONTACT PAD

TAPE ROLL

FOIL CONTACT PADS

16

TWO-WAY SWITCHES

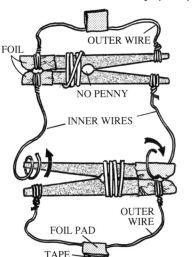

1. Replace the springs in 2 clothespins with rubber bands.

2. Wrap each jaw, "uppers" and "lowers," in pieces of foil. These jaws should *break* contact when the *wings* are squeezed together.

3. Point the jaws in opposite directions, then join the clothespins with 2 inner wires as shown.

4. Add an outer wire to each clothespin. The *wings* should *make* contact when squeezed together.

5. Attach foil contact pads with rolled tape at the middle of each outer wire. Write your name on your switch.

6. Wire your 2-way switch into a series circuit with a bulb and 2 dry cells.
   a. Diagram how your circuit works.
   b. Why are these switches ideal for a stairway?
   c. Your switch stays normally open. How could you modify it to stay closed?

17

SWITCHING MAZE

1. Connect 2 single switches in parallel. Connect these to 1 terminal of your double throw switch. Connect a 2-way set of switches to the other terminal. Tie in 2 cells and a bulb like this:

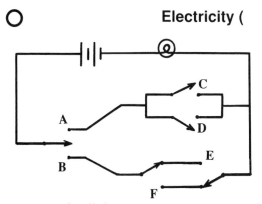

2. List 4 different switching pairs that operate the light.

3. Write a single "and/or" sentence that fully summarizes how to light the bulb.

4. Straighten a paper clip into a single wire. What wires could you interconnect so that…
   a. Switch A alone will operate the bulb?
   b. Switch B alone will operate the bulb?

18

BUILD A GALVANOMETER   Electricity ( )

1. Coil 3 m of thin insulated wire around the neck of a small jar. Twist and tape the leads to the side of the jar and write your name on it.

2. Scrape the insulation off the end of each wire. Loop the bare ends as usual, then attach foil contact pads.

3. Tape 7 cm of straight bare wire to a magnet as shown:

WIRE ARMS

4. Cut 2 notches in the end of a straw to fit over the magnet, and secure with tape. Slit a second straw along its length so it slides over the first.

SLIDING STRAW ➡ ⬅ NOTCHED STRAW

5. Rest the wire "arms" on the mouth of the jar. Adjust the wire "arms" and the sliding straw until it rocks back and forth about 1 swing per second.

1 SWING PER SECOND

6. Danish scientist Hans Christian Ørsted (1777-1851) made a wonderful discovery about electricity. Touch your galvanometer to a cell (only for an instant) and tell what he learned.

19

TEST YOUR GALVANOMETER ◯   Electricity ( )

RUN CURRENT BOTH WAYS

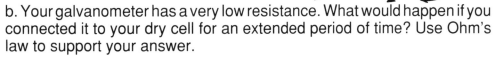

1. Run electricity through the coil of your galvanometer, first in one direction and then the other. (Touch the leads to your cell only for an *instant*.)

    a. Is your galvanometer sensitive to the direction of the current? Explain.

    b. Your galvanometer has a very low resistance. What would happen if you connected it to your dry cell for an extended period of time? Use Ohm's law to support your answer.

2. Adjust your galvanometer straw for maximum sensitivity: it should rock back and forth about once per second.

    a. See if you can detect current when 2 cells are connected in opposition. If both cells are equally matched, you'll need to watch closely.

    b. How can you decide which cell is stronger?

20

WHICH PATH? O   Electricity (   )

1. Wire 2 cells, 2 bulbs and 2 switches into a circuit like this:

2. How many bulbs light when you turn on…
   a. Switch x only?
   b. Switch y only?
   c. Both switch x and y?

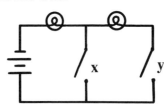

3. Apply Ohm's law to answer each questions:
   a. Why does 1 bulb shine brighter than 2?
   b. Why does only 1 bulb shine when both switches are closed?
   c. Might *some* current travel through the dark bulb when both switches are closed?

4. Test your hypothesis in step 3c: Wire your galvanometer in series, between switch y and its bulb. See if you can detect any current flowing through the dark bulb as you open and close switch y while keeping x closed.

21

VARIABLE RESISTOR O   Electricity (   )

1. Work in pairs. Pull off a single long strand from a ball of steel wool. Tape the ends across an index card.

2. Set up a variable resistance circuit like this, connecting all components in series.

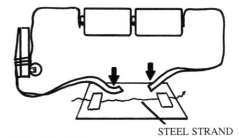

STEEL STRAND

   a. How do you change the brightness of the bulb?
   b. Explain your observations in terms of Ohm's law.

3. Add a second bulb to the circuit in parallel with the first.
   a. Explain how your variable resistor controls the brightness of both bulbs.
   b. It is said that electrons follow the path of least resistance. To what extent is this true?

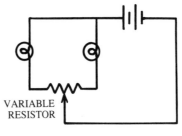

VARIABLE
RESISTOR

22

ADDING RESISTANCES **Electricity ( )**

1. Cut 2 index-card squares with sides the length of a paper clip. Then clip as shown.

2. Slide a medium and a thin strand of steel wool under the clips. Tape loose ends underneath.

MEDIUM STRAND   THIN STRAND

TAPE

3. Test *each* resistor with two cells and a bulb. Write your observations and conclusions.

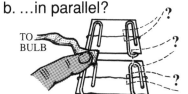

4. What happens to the total resistance when you add both resistors...

a. ...in series?

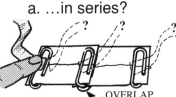

? ? ?

OVERLAP

b. ...in parallel?

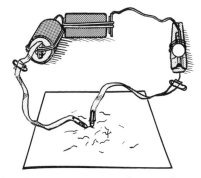

?

TO BULB

?

?

5. Resistors combine in series and parallel according to these formulas:

$$R_{series} = R_1 + R_2 + R_3 + ... \qquad 1/R_{parallel} = 1/R_1 + 1/R_2 + 1/R_3 + ...$$

a. If your thinner wire has a resistance of $3\,\Omega$ and your thicker wire $2\,\Omega$, find R_{series} and $R_{parallel}$.

b. Are your experimental results consistent with these formulas? Explain.

23

TINY FUSES **Electricity ( )**

1. Connect a foil ribbon at each end of a bulb and two cells in series. Attach a paper clip to each free end.

2. Gently pull on a piece of steel wool so iron fibers fall onto white scratch paper. Pass current through these tiny strands between the ends of your paper clips.

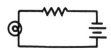

   a. Which fibers tend to melt and burn? Why?
   b. Which fibers tend to sustain the current and light the bulb? Why?

3. A fuse protects bulbs, dry cells, and other circuit components from short circuits. If the current surges, the fuse is the first to melt and break the circuit.
   a. Find a steel wool strand that is just the right size to make a good fuse. How should it perform?
   b. The cells in your circuit push about .25 amperes through the bulb. If you short the bulb, 3 to 15 times more current will surge through the circuit, depending on cell type and age. What size fuse (in amperes) is ideal for this circuit? Defend your answer.

24

ELECTROSCOPE (1)   O   Electricity (   )

1. Punch 2 holes in a piece of aluminum foil. Cut around the holes as close as you can to make 2 leaves like these.

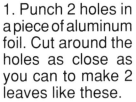

2. Cut a cardboard lid slightly larger than the mouth of the baby food jar. Stick a straight pin through the center, then bend up the point with pliers. Rest your leaves on this hook, and stick the lid on the jar with rolls of tape.

3. Charge a styrofoam cup by rubbing it with silk. Touch it and your finger to the scope.
   a. How does it show that it is charged?
   b. How can you quickly discharge it?

4. Try these ways to discharge (ground) a charged scope. Tell what happens and what you learn.
   a. Touch your finger to the edge of the lid.
   b. Try your pencil point and eraser.
   c. Try a penny on a thread; a can.
   d. Touch the scope with both feet off the ground. (Jump!)
   e. Breathe on the scope as if you were trying to fog a mirror.
   f. Don't do anything. Just observe the leaves for several minutes.

25

ELECTROSCOPE (2)   O   Electricity (   )

1. Diagram these different electroscope conditions using (+) and (-) symbols. Use $e^- \longrightarrow$ to show electron movement.
   a. (-) charged scope:
   b. (+) charged scope.
   c. (-) styrofoam nears a (-) scope.
   d. (+) silk nears a (-) scope.
   e. (-) styrofoam nears a (+) scope.
   f. (+) silk nears a (+) scope.

2. Pull up the pin head on your electroscope just a little, and twist a 30 cm wire around it like this:

3. Charge a styrofoam cup. Touch it to the wire so the leaves stay apart.
   a. Determine the charge on the leaves.
   b. Explain how the electroscope was charged *by contact*.

4. Charge your scope as you did in the previous activity. (Touch both your finger and the cup to the scope together.)
   a. Is your scope charged the same as in step 3?
   b. Propose a theory to explain your observations.
   c. Can you charge your scope *by induction*, without directly touching it with the negative cup? Explain.

26

ELECTROLYSIS (1) ◯ Electricity (   )

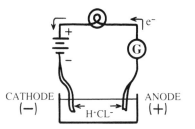

1. Connect 2 dry cells, a bulb and a galvanometer in series with foil ribbons attached to each end.

2. Pour a 5% solution of hydrochloric acid into a jar or beaker, then dip the end of each ribbon in opposite sides of the jar so they don't touch. Write your observations. There is much to notice!

3. Hydrochloric acid forms a solution of positive hydrogen ions (H^+) and negative chlorine ions (Cl^-). These ions receive or give up electrons to form hydrogen gas (H_2) and chlorine gas (Cl_2).

a. What reaction must be occurring at the negative electrode (the cathode)? Explain.

b. What reaction must be occurring at the positive electrode (the anode)? Explain.

4. Is electricity best described as a "flow of electrons" or a "flow of charge?" Explain.

5. Diagram how you might collect the gases that bubble off each electrode. (Don't actually do the experiment – chlorine gas is toxic!)

27

ELECTROLYSIS (2) ◯ Electricity (   )

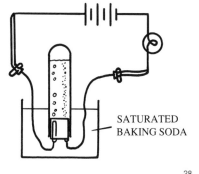

1. Cut a rectangle 2 paper clips long and 1 clip wide from an index card. Roll it the long way around a pencil, so it fits into the mouth of a *small* test tube.

2. Twist tie 30 cm wires to 2 paper clips. Clip these to opposite sides of the rolled card, then push them just into the test tube.

3. Fill the test tube *and* a jar or beaker with a saturated solution of baking soda. Close the test tube with your thumb, and invert it into the jar without letting air inside.

4. Wire this to a bulb plus 3 cells in series. If the bulb lights, your lead wires are shorting out and must be separated.

5. Collect gas until the water level drops to the top of the paper clip electrodes (about 5 minutes). Record your observations.

(Keep the tube inverted in the solution until the next activity.)

28

ELECTROLYSIS (3) O   Electricity (   )

1. Electrical energy breaks water apart into 2 gases. What are these gases and which gas forms at each electrode?

2. You can harmlessly burn the gas collected in the inverted tube. To do this:

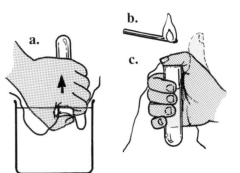

    a. Cover the mouth of the test tube with your finger while it is still submerged.
    b. Continue holding the tube upside-down while a friend strikes a match.
    c. Turn over the test tube, then uncover it directly *under* the match flame.
    d. Write your observations.

3. The gases in the test tube recombined to form water: $2H_2 + O_2 \rightarrow 2H_2O + energy$.

    a. Do hydrogen and oxygen contain more energy as separate gases, or as water?
    b. Energy can neither be created nor destroyed. Where did the energy come from to make your mini-explosion?

29

BUILD A WET CELL O   Electricity (   )

1. Roll a 3 x 6 cm piece of paper towel around the shaft of a galvanized nail. Leave its head and "neck" exposed.

2. Wind it from the neck down with about 1.5 m of bare copper wire in a single even layer. Wrap in the same direction as the towel is wrapped. *Don't let the wire touch the nail.*

3. Leave about a 6 cm lead at the bottom. Make another lead at the top by wrapping a second wire 2 or 3 times around the metal neck. *This top lead must not touch the wrapped wire.*

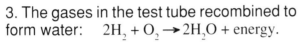

SECOND WIRE
WIRES MUST **NOT** TOUCH
FIRST WIRE
6 cm LEAD
TOWEL

4. Set your nail on a plate or petrie dish and soak it with 5% hydrochloric acid. Then drip 4 drops of hydrogen peroxide directly on the nail.

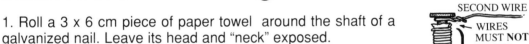

PEROXIDE
ACID
(G)

    a. Connect this wet cell to your galvanometer. Can you detect electricity?
    b. In which direction does the current flow? Explain how you know.

5. Acid in the towel dissolves zinc atoms on the galvanized nail, changing them to zinc ions and giving up electrons: $Zn \rightarrow Zn^{++} + 2e^-$. Meanwhile, hydrogen gas bubbles off the copper wire: $2H^+ + 2e^- \rightarrow H_2\uparrow$.

    a. Diagram how your cell works.   b. Compare this chemical cell to electrolysis.

30

BUILD A STORAGE CELL ◯ Electricity ( )

1. Fix several lead fishing sinkers to the ends of two pieces of wire. (You may need pliers.) Hang them over the rim of a small jar filled with saturated salt water. Connect each wire to a foil ribbon.

2. Wire this to 2 cells in series, your double throw switch and your galvanometer.

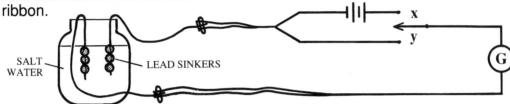

SALT WATER — LEAD SINKERS

x
y
G

3. Charge your lead storage cell for about 30 seconds. (Close the x side.)
   a. Note which way your galvanometer deflects.
   b. Note any evidence of a chemical reaction.

4. Steady the galvanometer needle so it is nearly motionless, then discharge your storage cell through the galvanometer. (Close the y side.)
   a. Does the current still flow in the same direction? Explain.
   b. Summarize how your storage cell converts chemical and electrical energy.

31

CHARGE A CAPACITOR ◯ Electricity ( )

1. Get a capacitor and a commercial galvanometer. Connect these to a dry cell and a 2-way switch like this:

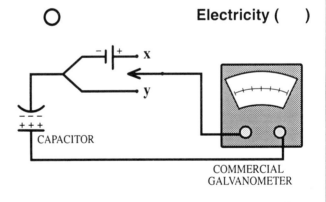

x
y

CAPACITOR

COMMERCIAL GALVANOMETER

2. Capacitors have plates that store electric charge. Negative charge builds up on one plate while positive charge builds up on the other.

   a. How do you operate the switch to charge and discharge these plates? How does your galvanometer respond?
   b. Compare this capacitor to the storage cell you made previously.

3. Fully charge your capacitor. Devise a way to confirm that it really does contain a full charge without discharging it.

4. Explain how to give your capacitor a partial charge. How can you tell it's neither "full" nor "empty?"

5. Does your capacitor "leak" charge over time? How do you know?

32

AMMETER (1) ◯   Electricity (   )

1. Lightly tape a narrow foil pointer to the top of your galvanometer. Adjust the straw to swing from side to side in about 1 second.

2. Put a lump of clay on the top of an inverted baby food jar and stick in a straw. Cut out a paper ruler and fix it to the top with tape rolled sticky side out. Adjust so its millimeter divisions form a scale just behind the freely swinging pointer.

3. Connect the galvanometer to your dry cell, ribbon and a 100 Ω resistor. Keep a second 100 Ω resistor nearby for testing.

4. Work in a space with little or no breeze. Record how many millimeters the pointer deflects when the current passes through 100 Ω. Repeat for 200 Ω, then 50 Ω. Write a full report that includes your data, methods, and conclusions.

33

AMMETER (2) ◯   Electricity (   )

1. Set up your galvanometer as before, with the millimeter scale resting just behind the pointer. This pointer should move freely, about one swing per second.

2. An average 1.5 volt dry cell connected to a 50 Ω resistance produces about 25 mA of current (.025 A). Through how many mm will this much current deflect the pointer?

3. Knowing how far each 25 mA of current deflects the straw, construct a scale up to 500 mA. Curve it to match the swing of your pointer. (Keep your galvanometer in a safe place so it remains at this setting.)

4. Attach this scale to the straw set in clay as before, so the zero point rests behind your galvanometer pointer. You have just made an ammeter!

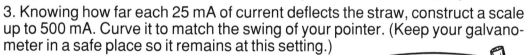

   a. Replace the 50 Ω resistors with your light bulb, and measure the current.
   b. How does the resistance of your bulb compare to a 50 Ω resistance?
   c. Momentarily short out your cell with no light bulb in between. What happens? Why?
   d. What is the difference between a galvanometer and an ammeter?

34

GENERATORS AND MOTORS   O   Electricity (   )

1. Change your galvanometer to a generator:
   a. Remove or fold down the magnet's wire arms. Hook the jar to a commercial galvanometer.
   b. Rest the straw on the jar's rim inside the coil, and rotate it in just one direction.
2. You have just produced electicity:
   a. How did the galvanometer needle move?
   b. What does this tell you about the flow of electricity?
   c. AC means alternating current. DC means direct current. Which kind of electricity does your generator make? Your chemical cell?

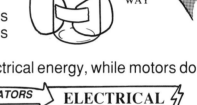

3. Generators change mechanical energy to electrical energy, while motors do the opposite.

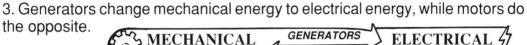

MECHANICAL ENERGY   *GENERATORS*   ELECTRICAL ENERGY   *MOTORS*

   a. Which is the generator in this experiment? Explain.
   b. Which is the motor in this experiment? Explain.

35

INTERNAL RESISTANCE   O   Electricity (   )

1. Fold a paper towel in half 3 times to form a narrow folded strip. Then draw a 24 cm ruler on its surface, with a penny circled at one end. Number it as shown:
2. Make a chemical cell: soak the strip in vinegar and lay it on a plate.

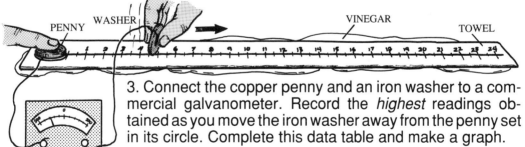

PENNY   WASHER   VINEGAR   TOWEL

3. Connect the copper penny and an iron washer to a commercial galvanometer. Record the *highest* readings obtained as you move the iron washer away from the penny set in its circle. Complete this data table and make a graph.

4. Explain how your graph supports Ohm's law. (Hint: The distance separating the coin from the washer is proportional to the cell's internal resistance.)

distance between electrodes (cm)	needle deflection (arbitrary units)
0	
1	
2	
3	

36

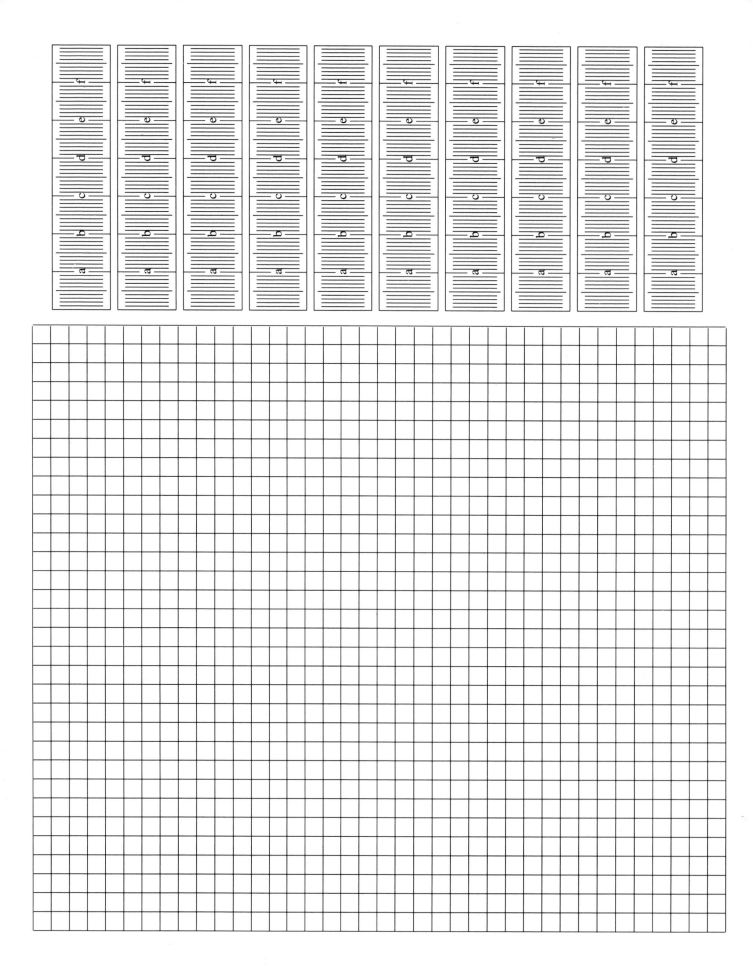